Springer Biographies

The books published in the Springer Biographies tell of the life and work of scholars, innovators, and pioneers in all fields of learning and throughout the ages. Prominent scientists and philosophers will feature, but so too will lesser known personalities whose significant contributions deserve greater recognition and whose remarkable life stories will stir and motivate readers. Authored by historians and other academic writers, the volumes describe and analyse the main achievements of their subjects in manner accessible to nonspecialists, interweaving these with salient aspects of the protagonists' personal lives. Autobiographies and memoirs also fall into the scope of the series.

More information about this series at http://www.springer.com/series/13617

Bo Lojek

William Shockley: The Will to Think

Bo Lojek
ATMEL Corporation
Colorado Springs
Colorado, USA

ISSN 2365-0613   ISSN 2365-0621   (electronic)
Springer Biographies
ISBN 978-3-030-65957-8   ISBN 978-3-030-65958-5   (eBook)
https://doi.org/10.1007/978-3-030-65958-5

This Springer imprint is published by the registered company Springer Nature Switzerland AG
The registered company address is: Gewerbestrasse 11, 6330 Cham, Switzerland

*To my Mom and my Dad who taught me that
life is a true test of intelligence*

Acknowledgements

Several people have contributed to this book. I am indebted to Atmel VP, Dan Malinaric, who for several decades shielded me from the communist invention of teamwork, and allowed me always to work as an individual contributor.

My most candid and trusted critic, Mary Martin, spent countless hours with me searching in archives across the country.

My friend and RTA cohort, Jeff Gelpey, provided many constructive suggestions and advice which improved the clarity of the text.

I belong to the older generation for whom, as someone once said, "*A library is not a luxury but one of the necessities of life.*" I was fortunate to receive help from librarians belonging to the library system of the University of Colorado. Don Pawl, Interlibrary Loan Librarian, was on many occasions able to obtain items from obscure sources located in generous libraries around the world.

Prof. Arthur Jensen kindly allowed me to study and use the collection of the Foundation for Research and Education on Eugenics and Dysgenics (FREED) documents.

The late Ian Ross shared with me details of Shockley's work on field effect devices and the booklet containing jokes Shockley collected while at Bell Laboratories. Unfortunately, most of them are not suitable for print.

My friends of several decades, the late Esther M. Conwell, Ray Warner, and Morgan and Betty Sparks, provided valuable insight into the Bell Laboratories culture and Shockley's personality.

The beauty of baroque music allows me to be more at ease with myself, so it is appropriate to express my thanks to François Couperin whose music is always in the background during writing.

I thank the editorial staff of Springer for their professional support and patience. Zach Evenson, Angela Lahee, and Stephen N. Lyle for their unwavering attention to consistency and style, which has been an example of excellence.

Finally, my gratitude goes to all my beloved Schnauzers who have guided me through my entire life. The serial number four, Filip von Langdale IV, always knows the right time to pull me away from my desk and take me for a walk in the woods.

Contents

Chapter 1
History: An Engineering View

"History is a chronological record of significant events. An historian is a person who is an authority on history and who studies it and writes about it; a chronicler; an analyst, aware of their inability ever to reconstruct a dead world in its completeness."

"Engineering is the discipline dealing with the art and science of applying scientific knowledge to practical problems. Practitioners of engineering are called engineers."

[1913, Webster]

1.1 History: An Engineering View

The modern "wireless" society, where it is almost impossible to stay still and ponder for more than a few seconds, idolizes heroes. For this reason, discoveries of radical inventions are always glorified and given an aura of mystery. Many people still have an idealized picture of the lone inventor in a laboratory, hidden away from the outside world for many years, awaiting his moment of glory. In reality, the lone inventor is rather the exception than the rule. Although the lone inventor still exists, the vast majority of innovation is the result of the work of many individuals, each adding some separate component to the final solution. The label "hero" is then assigned to the individual who contributed the final component leading to a radical innovation. All other contributors and the components they added, although crucial to the final solution, are then forgotten.

Another myth is that radical inventions are always based on completely new knowledge. In fact, in the vast majority of cases, it is some unconventional combination of existing knowledge that is the ultimate source of novelty. Radical inventions are only rarely based on completely new knowledge. Frequently, even a simple rearrangement of facts that are already common knowledge can be the main source for a radical invention.

In this sense, a history of engineering might be viewed as a series of incremental technical changes. Most of these changes can be characterized as incremental improvements with limited impact on the economic system. Occasionally, the change might result in a radical or breakthrough invention. Radical inventions are those inventions that serve as a source for subsequent inventions, and they are frequently viewed at the time of conception as being a risky departure from existing practice. Successful radical inventions tend to provide an opportunity for the inventing firm to

© The Author(s), under exclusive license to Springer Nature Switzerland AG 2021
B. Lojek, *William Shockley: The Will to Think*, Springer Biographies,
https://doi.org/10.1007/978-3-030-65958-5_1

gain a sustainable competitive advantage, with a consequent generation of economic profit. The reality of life, however, demonstrates that top-level managers often lack a deep enough understanding of emerging technologies to be able to develop radical inventions. In contrast, mature technologies are well understood and have been tested and used in many different settings. For this reason, they offer little risk and much greater reliability relative to newly developed and less well-tested technologies. It is safer to prefer mature technologies to nascent technologies. The outcomes of emerging technologies are much less certain, so radical innovations are not always welcome. They thus very seldom result from organized and managed effort. One rather atypical exception was the research carried out at the Bell Telephone Laboratories, which generated many Nobel Prizes and patents, and made major contributions to the information age before its demise in the early 1980s.

There are also cases of radical inventions which are not always publicly known and are sometimes forgotten, not because they are not useful, but because at the time of their development they were so advanced and ahead of their time that they could not be used. When Ferdinand Braun discovered the rectifying effect there was no application for it.

Often the name of the inventor to whom an invention is attributed varies from country to country, depending on the country of origin of the authors. A typical example is the invention of the telephone. Alexander Graham Bell, who filed his patent application "Improvement of Telegraphy" on February 14, 1876 is almost universally recognized as the inventor of the first telephone. If we omit the controversy over whether Elisha Gray's patent application arrived before or after Bell's submission, the historical evidence regarding this invention points to Johan Phillip Reis (1834–1874). Reis imagined that electricity could be propagated through space, as light can, without the aid of a material conductor, and he performed some experiments on the subject. The results were described in a paper, "On the Radiation of Electricity", which, in 1859, he mailed to Professor Poggendorff for publication in the Annalen der Physik. The manuscript was rejected. Reis continued in his work and on October 20, 1861 presented a seminar "On the propagation of tones over arbitrary distances via galvanic currents" in Frankfurt/Main. He demonstrated his apparatus by transmission of the sentence: *"The horse does not eat cucumber salad"*. The first prototype of an instrument could transmit a signal over a distance of 100 meters. In 1862, he again tried to interest Poggendorff with an account of his instrument, referring to it for the first time as *"die telephone"*. His second offering was rejected like the first, as the editor considered Reis' invention of the transmission of speech by electricity as a "chimera". The Physical Society of Frankfurt rejected the apparatus and saw the instrument as a mere "philosophical toy". But Reis believed in his invention, even if no one else did. He continued his lectures even though he had been stricken with tuberculosis. When he gave a lecture on the telephone at Gießen in 1864, Poggendorff, who was present, invited him to send a description of his instrument to the Annalen. Reis replied: *"Ich danke Ihnen sehr, Herr Professor, aber es ist zu spät. Jetzt will ich ihn nicht schicken. Mein Apparat wird ohne Beschreibung in den Annalen bekannt warden"* (Fig. 1.1)[1].

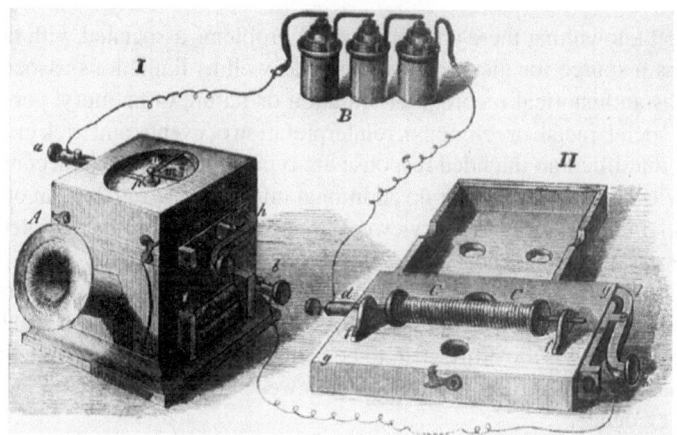

Fig. 1.1 The Reis "telephone" (1861)

Inventions like Ohl's p-n junction, Hoerni's planar process, Frohman-Bentchkowsky's non-volatile memory cell, Craford's yellow LED to name a few, are typical examples illustrating that the personality traits of the contributor are what is of the greatest importance. The common feature of all these inventors was their individualism. Personal individualism and critical thought give people the ability to be creative, while teamwork tends to destroy creativity; team spirit often inhibit thinking and is a perfect hideaway for incompetent members of the team. Every single innovation described in this book is the result of the individual, not the team.

Personal individualism, however, might lead to an extreme form of individualism, better known as egotism. Although almost all contributors to radical inventions tend to be of generous temperament, once a certain innovation becomes a success and looks like it may make history, it is only natural that the egotist should want to be a part of it.

The problem is that such an individual could be someone who has had very little to do with the struggles and hard work that have eventually led to the radical innovation. This is one of the reasons why interpretations of recent history, in which some of the participants of the event are still around, are subject to constant controversy.

It is natural that nobody should want to be associated with failure, while everyone is interested in being part of success. When I was doing research for my previous book, I did not find any transcript or recording of an oral interview in which the interviewee acknowledged that he had made or participated in a wrong or incompetent decision. The reader should then ask the following question: Why are companies like Fairchild,

---

[1] Mr. Professor, thank you very much, but it is too late. I do not want to submit it now. My device will be known even without description in the Annalen.

RCA, Westinghouse, or Motorola, to name but a few, no longer in business, when they were managed by such exceptional individuals as those in the oral interviews would claim?

It is well known that there are a number of problems associated with taking oral histories as a source for historical evidence, as well as limitations associated with their use as an historical record. Deterioration or failure of memory, personal bias (political, social, racial, or religious), reinterpretation of events, and trick or confusing questions that illicit an intended response are typical problems that accompany any oral interview. If there is little or no additional information available, an oral history can only claim to present the interviewee's interpretation, and such an interpretation cannot necessarily be taken as historical evidence.

When the right time comes, the individual's interests usually shift to a second major goal in life, which is worldly success, with its three prongs of wealth, fame, and power. This too is a worthy goal, to be neither scorned nor condemned; the only issue is that one needs to confront the interviewee's interpretation with indisputable historical evidence.

I have often been puzzled as to how and why historians assigned a particular invention to a particular person. Since history deals with people and events of long ago, how do we know if it can be trusted? History is full of stories created by the winners, and by those with a vested interest in one side or the other. Of course, the losers are and were always the bad guys. Everyone is free to examine the past and form their own conclusions. But it has one significant disadvantage: "popular history" and what really happened are rarely the same.

As an engineer, I was always interested not only in the history of science but also in the technical details involved in the invention. Once you understand the solution behind the invention, it is not difficult to recognize what kind of previous efforts led to the invention. Being familiar with history can open minds to discoveries, fascinating people, and different ways of looking at things. If we approach history from this point of view, there are, of course, some historical "facts" that are not and cannot be in dispute.

1.2   About This Book

> *"The profoundest of all infidelities is the fear that true will be bad."*
>
> Herbert Spencer (1820–1903)

This book *"William B. Shockley: The will to think"* describes the life and accomplishments of the Nobel laureate William Bradford Shockley in a way that standard biographies often neglect in favor of sensational descriptions of his views of the social problems of his time or salacious lies about his personal life. Although the title might not be specific enough, the book is written for people with an engineering attitude who want to understand where we were and why we got to where we are.

The reader will find that, in the majority of cases, this book differs from common folklore and "ideologically correct" science as portrayed in numerous "fashionable" oral and written histories shared by various institutions and individuals in several publications and especially on the internet. This book may appear superfluous to those readers who purport to already know everything or who are *dramatis personae*. If you fall into this category of readers, this book may hurt your feelings.

William B. Shockley was a very complicated and difficult person to understand. In the first half of his life as a physicist, he frightened with his brilliance. In the second half of his life, he delved into several taboo subjects. It is even more complicated to put such a subject in an historical perspective as we all wish to avoid having a cloud of prejudice hanging over us.

When we meet a difficult person like Shockley, our instinct is to try to change them. Several of Shockley's friends tried but it never worked! The only way to disengage a difficult person is to try to understand where they are coming from. Try to find what drives their decisions. For some people it is money; for others, it is power. For Shockley, it was the search for scientific truth.

The author of this book is not an historian, but rather a witness. In this book I employ a process of reasoning to determine the factual information. But the social sciences Shockley touched upon are not exact sciences, as we in the realms of engineering and physics know them. There are always the issues of opinion and interpretation. A reader may ask "If this book does not support my world view, then why should I read it?" You may have already formed your own opinion about Shockley and his beliefs, and that may be all that is important to you. Then there is no way that you can be enlightened by this book. Should you continue reading, I am not asking you to accept my opinion. Rather, I invite you to create your own by precision reading, comparing the timeline, and studying the enclosed historical documents.

When a text is considered important, it becomes a critical issue to read it properly and understand it appropriately. This is perhaps one of the most comprehensive ways you can learn about a topic. Almost without fail, the more you read on a topic, the better the understanding you will eventually have of it. Sometimes you might be surprised by elements that you previously overlooked or had not even considered. You may find that the topic is not necessarily black and white and perhaps there are times when you may make an exception to your beliefs.

My students repeatedly tell me "Reading books is a waste of time. All the knowledge you need is online." I do not find Internet to be a reliable source of information and, more importantly, the computer screen constantly diverts the reader's attention by messages not related to the subject of study. Good books make for better sources of knowledge. Longer articles seem to be better read in print. Research on memory tells us that we learn by connecting new information to previous knowledge. Flipping back and forth to connect sections to each other may help us make those connections in the brain more quickly and easily.

The point is intelligent discovery of what is true. If there is nothing to discover that is true then there is no reason discussing, disputing, arguing, making an issue of anything, because there is nothing to believe in. When you read a book you might disagree with, you learn. You may discover one thing or many things, but you will learn something. And it will help you become a more informed person.

Chapter 2
Prologue: The Enigma of Shockley

> *"What you are trying to do seems to be absolutely essential to the future of the world but I guess that you will get very little thanks for it. I hope your courage would hold up."*
>
> Dr. John J. Osborn
> (letter to W. Shockley, April 18, 1966)

The highest aim of science should be the ultimate search for truth. Today we admire scientists like Galileo Galilei for their intellectual honesty about what they saw, even if the results were uncomfortable to political establishment. Hungarian physician Ignaz Semmelweis was committed to an insane asylum after losing his job for suggesting, in the nineteenth century, the radical notion that infections could be spread by germs on doctors' hands in hospitals. In 1600, philosopher Giordano Bruno was burned at the stake for the heresy of proposing that the universe might be infinite. Dr. Shockley was such a hero who, in the search for true science, paid no attention to moral convention and feelings.

The father of the transistor, and arguably the most influential inventor of the last century, William Shockley was the leader of the team that created the seminal invention of the century. There are those who were offended by his abrasive personality and politically incorrect views, who minimize his role in inventing the transistor. He was the father of Silicon Valley; his company the originator from which virtually all the Valley's dominant companies and technologies would emerge. Modern microelectronics contains the technical descendants of Shockley's work.

He and his friend, James B. Fisk, designed a nuclear reactor several years before the Manhattan Project scientists at Los Alamos. In 1939, much of the physics community was taken by the growing advances toward fission made by European scientists. Shockley and James Fisk were assigned by the Bell Laboratories to examine the potential for fission as an energy source. Shockley came up with an idea: *"if you put the uranium in chunks, separated lumps or something, the neutrons might be able to slow down and not get captured and then be able to hit the U-235."* In a few months, he and Fisk designed one of the world's first nuclear reactors. Their report went immediately to Washington. The government classified it right away, even keeping it secret from its own scientists. The authorities fought any attempt by Fisk, Shockley, or the labs to take out a patent.

 B. Lojek, *William Shockley: The Will to Think*, Springer Biographies, https://doi.org/10.1007/978-3-030-65958-5_2

Shockley may have saved thousands of lives without leaving his desk. When war broke out, P. Morse was recruited to research munitions problems the Navy was having, mostly with its depth charges. Shockley volunteered to join Morse's office, the Anti-Submarine Warfare Operations Group. Under Morse's guidance, Shockley and his team solved the depth charge problem and successful attacks on German U-boats increased by a factor of five. Shockley's main weapon was the science of operations research, then largely ignored in the U.S., but already recruited for the war effort by the British. He then went about changing the way the Navy searched for submarines, again improving the kill-ratio. He devised tactics for the Atlantic convoys to evade German bombers after determining statistically—and without ever seeing either a convoy or a bomber—that the bombers did not carry radar. Shockley eventually wound up in the Army Air Corps. helping train bomber crews in the European theater. He was a leading proponent of the science of operations research in America, beginning in World War II, with desk-bound calculations that probably saved tens of thousands of lives. Although he won the highest possible civilian honor for his work, that work has long been forgotten.

The spotlight was turned upon him later when he became involved in a controversial topic in which he became avidly interested: the genetic basis of intelligence. During the 1960s, he argued, in a series of articles and speeches, that people of African descent have a genetically inferior mental capacity when compared to those with Caucasian ancestry. This hypothesis became the subject of intense and acrimonious debate. The press coverage ignored the scientific basis and data of Shockley's arguments and frequently referred to Shockley's view as "race prejudice".

Shockley did not know what needed to be done, but he thought that first we needed to find out what was the root cause of what he called the heredity-poverty-crime problem and then find a remedy for how to improve this unsatisfactory situation. He sent letters to members of the National Academy of Sciences asking for their support. Shockley contacted his friend, Frederick Seitz, who was president of the National Academy of Sciences, and asked him for help. He urged "*do the research, find facts and discuss them widely.*" Shockley stated in the letter to F. Seitz: "*My position is not that all Negroes are inferior to all whites: I do believe that many Negroes are superior to many whites, in fact my statistical studies show that American Negroes achieve almost every eminent distinction that whites achieve.*"

Shockley in his presentation "*A try simplest cases approach to the heredity-poverty-crime problem*" read before the Academy on April 26, 1967 stated: "*What can be done to make a diagnosis? I have two recommendations: First, I believe that a National Study Group should be set up to do research and find out definitive conclusions. Second, a study of drastic changes in environment on the most disadvantaged children should be taken.*"

F. Seitz in a letter to Shockley dated July 22, 1966 stated "*I can think of few problems more sticky than trying to decide further what can be done about them*" (Fig. 2.1).

Shockley, motivated by repulsion of his peers at the National Academy of Sciences, and using data taken primarily from U.S. Army IQ tests and from the U.S. Office of Education, drew the conclusion that the genetic component of a person's

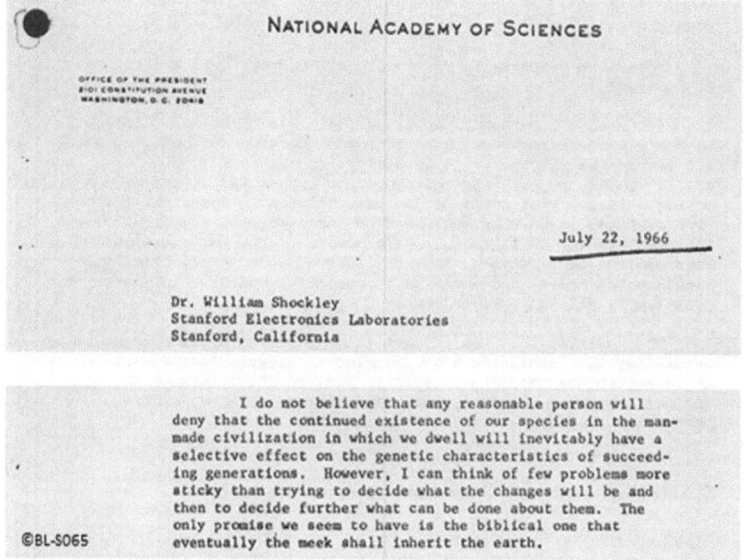

Fig. 2.1 Seitz letter to W. Shockley (July 22, 1966)

intelligence was based on genetic heritage. A similar hypothesis was earlier advanced by James Watson and Francis Crick, who in 1953 marked a milestone in the history of science and gave rise to modern molecular biology, a discipline largely concerned with understanding how genes control the chemical processes involved in copying genetic material. Their model enabled explanation of the molecular structure of nucleic acids, and also pinpointed DNA as the carrier of genetic information.

Textbooks on molecular biology or molecular genetics emphasize[1,2] *"with the exception of some viruses, almost all organisms on this planet store their cellular blueprints for life in double stranded DNA molecules."* Amongst eminent psychologists and behavioral geneticists,[3,4] it is a nearly incontrovertible fact that intelligence is highly heritable and one of the single best predictors of long-term educational and occupational success, lending modern-day credence to Shockley's conclusions regarding the heritability of intelligence.

Although Shockley made the data he used and their statistical analysis public, up to now no one has offered an earnest rebuke of the Shockley data pointing out fundamental errors in his statistical analysis. In fact, some of the latest data collected

---

[1] R. L. Miesfeld, *"Applied Molecular Genetics"*, J. Wiley & Sons, NY 1999.

[2] O. Brandenberg et al., *"Introduction to Molecular Biology and Genetic Engineering"*, UN Rome 2011.

[3] R. Plomin, S. von Stumm, *"The new genetics of intelligence"*, Nature Reviews Genetics, Vol 19 (2018), pp. 148–159.

[4] I. J. Deary, W. Johnson, L.M. Houlihan, "Genetic foundation of human intelligence", Human Genetics, Vol. 126 (2009), pp. 215–232.

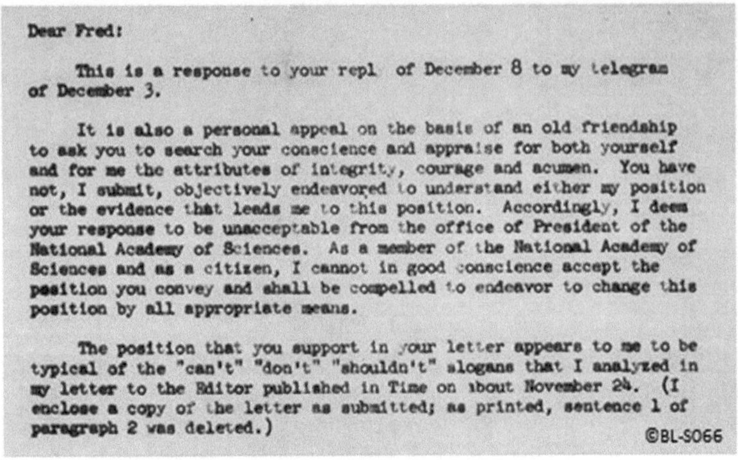

Dear Fred:

This is a response to your repl. of December 8 to my telegram of December 3.

It is also a personal appeal on the basis of an old friendship to ask you to search your conscience and appraise for both yourself and for me the attributes of integrity, courage and acumen. You have not, I submit, objectively endeavored to understand either my position or the evidence that leads me to this position. Accordingly, I deem your response to be unacceptable from the office of President of the National Academy of Sciences. As a member of the National Academy of Sciences and as a citizen, I cannot in good conscience accept the position you convey and shall be compelled to endeavor to change this position by all appropriate means.

The position that you support in your letter appears to me to be typical of the "can't" "don't" "shouldn't" slogans that I analyzed in my letter to the Editor published in Time on about November 24. (I enclose a copy of the letter as submitted; as printed, sentence 1 of paragraph 2 was deleted.)

©BL-S066

Fig. 2.2 W. Shockley's reply to F. Seitz dated December 15, 1956

by the U.S. Department of Education still report[5,6] the same trend in the gap in educational achievements as Shockley's analysis concluded some fifty years ago. It is interesting to note that these disparities have persisted over a span of some five decades, a period which has seen rapid and extensive societal, economic, and technological change both in the US and across the globe. Yet, despite all of the aforementioned progress, it seems that little has been accomplished to effectively close the achievement gap.

It is worth pointing out the obvious emotion-laden nature of this highly controversial issue, oftentimes preventing a purely objective and rational look at the data, from either end of the ideological spectrum. For many, sadly, racial differences are unpleasant matters that should not be discussed in polite society and, if ignored, might hopefully disappear. Even in Shockley's day, various political and ideological pressures aroused pushback on his proposed research studies.

Leveraging his personal friendship with Academy president F. Seitz, Shockley repeatedly provided him with documents relevant to his concerns and asked for research to be carried out to find out the root cause of these problems (Fig. 2.2).

Seitz delayed his replies to Shockley's letters or did not reply at all. Finally, on January 8, 1968 Seitz sent Shockley a letter explaining that because "…the American Negro tends to live within a social framework different from his white counterpart … there is probably no significant role for truly scientific study" (Fig. 2.3).

---

[5] Achievement Gaps: How Black and White Students in Public Schools Perform in Mathematics and Reading on the National Assessment of Educational Progress Statistical Analysis Report, U.S. Department of Education NCES 2009-455, July 2009.

[6] School Composition and the Black-White Achievement Gap, National Center for Education Statistics, U.S. Department of Education, September 2015.

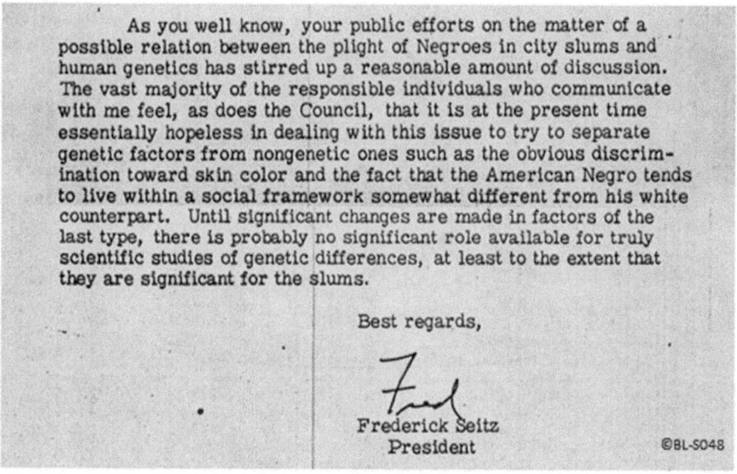

> As you well know, your public efforts on the matter of a possible relation between the plight of Negroes in city slums and human genetics has stirred up a reasonable amount of discussion. The vast majority of the responsible individuals who communicate with me feel, as does the Council, that it is at the present time essentially hopeless in dealing with this issue to try to separate genetic factors from nongenetic ones such as the obvious discrimination toward skin color and the fact that the American Negro tends to live within a social framework somewhat different from his white counterpart. Until significant changes are made in factors of the last type, there is probably no significant role available for truly scientific studies of genetic differences, at least to the extent that they are significant for the slums.
>
> Best regards,
>
> Frederick Seitz
> President   ©BL-S048

Fig. 2.3 F. Seitz's last letter to W. Shockley (January 8, 1968)

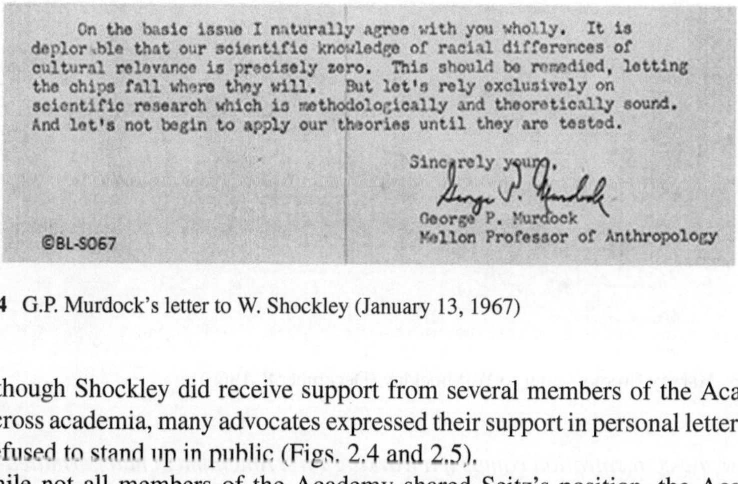

> On the basic issue I naturally agree with you wholly. It is deplorable that our scientific knowledge of racial differences of cultural relevance is precisely zero. This should be remedied, letting the chips fall where they will. But let's rely exclusively on scientific research which is methodologically and theoretically sound. And let's not begin to apply our theories until they are tested.
>
> Sincerely yours,
>
> George P. Murdock
> Mellon Professor of Anthropology
>
> ©BL-S067

Fig. 2.4 G.P. Murdock's letter to W. Shockley (January 13, 1967)

Although Shockley did receive support from several members of the Academy and across academia, many advocates expressed their support in personal letters only and refused to stand up in public (Figs. 2.4 and 2.5).

While not all members of the Academy shared Seitz's position, the Academy Social Science Research Council was, in the end, loyal to its President, and Shockley's proposal for a research study was repeatedly rejected.

Shockley repeated his appeal again on April 28, 1969 in a scientific presentation "A Polymolecular Interpretation of Growth Rates of Social Problems" accompanied by charts and statistical data. In this presentation he also stated: "*Eugenics is a shunned word because it was a feature of Nazi-Aryan supremacy. But the lesson of Nazi history is not that eugenics is intolerable. One hundred and forty years before Hitler, the lesson to be learned from Nazi history was incorporated into our Constitution as the First Amendment guaranteeing freedom of speech and of the press. Only the most anti-Teutonic racist can believe that the German public would*

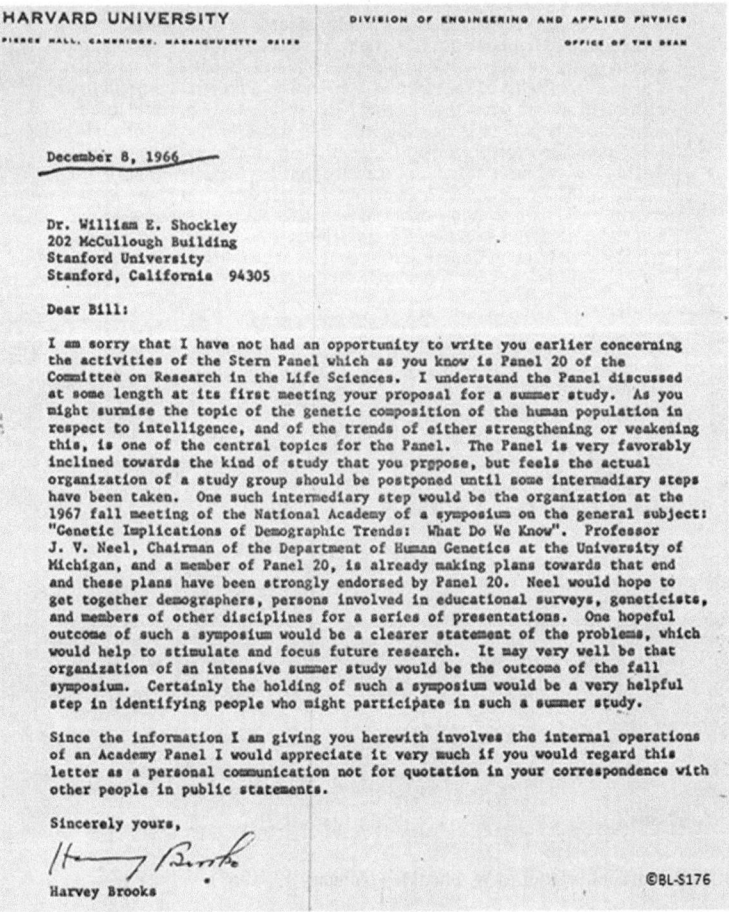

Fig. 2.5 Harvey Brooks' letter to W. Shockley (December 8, 1966)

tolerate the concentration camps if a working First Amendment had permitted public exposure and discussion of the genocide".

The Harvard geneticist David Reich[7] emphasizes the arbitrary nature of traditional racial groupings, but still argues that long periods of ancestry on separate continents have left their genetic marks on modern populations. These are most evident for physical traits like skin and hair color, where genetic causation is entirely uncontroversial. However, Reich asserts that all genetic traits, including those that affect behavior and cognition, are expected to differ between populations or races. To overemphasize the genetic factors, you may ask yourself the question in which

---

[7]D. Reich, *"How Genetics is Changing Our Understanding of 'Race'"*, The New York Times, March 23, 1918.

A POLYMOLECULAR INTERPRETATION OF
GROWTH RATES OF SOCIAL PROBLEMS

By W. Shockley
Stanford University

(To be read in part Monday, 28 Apr 69 PM at
National Academy of Sciences)

I   DYSGENICS AND EUGENICS HISTORY

"Dysgenic" is so fancy and abstruse an adjective that I have been
warned that if I use it, I shall be ignored. Yet it is the unintended dysgenic
byproducts of present social welfare programs that I and many thoughtful
and responsible citizens fear may pose a major, if not the major, threat to
our nation's next generation of citizens. I believe dysgenic ailments must
be diagnosed and, if necessary, cured.

RECOMMENDATION

We urge the public, the press, the government, and the scientific

community to seek facts relevant to hereditary aspects of our national

human quality problems. We believe that from such inquiry will inevitably

come knowledge suggesting wise, humane and appropriate remedial

legislation.
                                                                    ©BL-S064

Fig. 2.6 Transcript of Shockley's presentation at NAS (April 28, 1969)

environment you need to live if you want to change your eyes or hair color. What environment can change our mental abilities?

Pinker wrote[8] *"The profound questions are about what, precisely, are the non-genetic causes of personality and intelligence."* Unfortunately, the public does not get most of its information about genetics from molecular biologists but instead from popular media with no or vague definitions of the technical words used in arguments and discussions amongst experts. Many popular media authors are, in fact, unaware of theoretical advances in the field, long after the new way of thinking has become common in the field. To answer Pinker's question, thus, requires not only overcoming ideological biases, but also an insistence that media professionals act as facilitators—not interpreters—of scientific evidence to the general public. This, however, can only be achieved through thought, or rather, a will to think.

Shockley described an early meeting with Enrico Fermi, the wartime A-bomb physicist. Fermi said that one of the most important things is the *"will to think."* Shockley wrote later: *"A competent thinker will be reluctant to commit himself to the effort that tedious and precise thinking demands—he will lack "the will to think"*

---

[8]S. Pinker, *"Why Nature & Nurture won't go away"*, Daedalus Vol. 133 (2004), pp. 5–17.

unless he has the conviction that something worthwhile will be done with the results of his efforts—and, of course, there is always also the risk that his hard thinking may not produce any creative ideas". Thinking is one of the most difficult things a man can do. People understand investing time for farming and gardening that yield harvests months later. People readily invest money in a bank that yields interest years later. But people resist investing time in thinking because, unlike farming and banking, rewards in harvesting knowledge are delayed; there is no immediate or predictable payback.

Shockley's unusual personality had a major effect in shaping the personality of Bell Laboratories, and by extension, that of the microelectronics industry. In later years he fell from grace because of his views on the genetic basis of intelligence. Shockley argued that the higher rate of reproduction among the less intelligent had a dysgenic effect that would ultimately lead to a decline of civilization. The issue with William Shockley is that his scientific achievements outweigh by far any of his views we might see as objectionable. On the 50th anniversary of the invention of the transistor Isaac Asimov called Shockley's junction transistor "*perhaps the most astonishing revolution of all the scientific revolutions that have taken place in human history.*"

I met William Shockley and his wife, Emmy, for the first time as a boy. He was a wise and decorated man, and to me, of course, seemingly an old man. Shockley's teaching methods were severe, sometimes brutal. Equally severe and ruthless was his criticism. But I was fascinated with the exquisite way he explained complicated problems and with his specific humor. He was very bright and truly ingenious, with a quick grasp of new ideas.

I have passed that age now and I cannot escape the desire to see him again. Once, Shockley told me "*you do not need to agree with me, just admit for a moment, what if this is true*". I often ponder about his statement and I ask myself the question "*Is knowledge of certain kinds dangerous or undesirable*"*? Can certain knowledge hurt us?* If you continue reading the following chapters, ask yourself questions, confront the time sequence of events, check the reproduced documents and imagine "what if". The world can be made better by knowledge, not by ignorance.

Chapter 3
Mother and Father

"The true Genius is a mind of large general powers, accidentally determined to some particular direction."

Samuel Johnson
(author of the first English dictionary).

The word "genius" derives from the same roots as "gene" and "genetic" and meant a tutelary god or spirit given to a person.

William Bradford Shockley, a genius of the kind that appears on the Earth only every few centuries, was born on Sunday, February 13, 1910 at 69 Victoria St., London. His independent and free-thinking American born parents, May Bradford and William Hillman Shockley, married in 1908, and lived in London from 1909 to 1913.

May was born as Cora May Wheeler on May 11, 1879 in Moberly, Missouri to Sallie Jane Leona Ward who divorced May's father and remarried the mining surveyor Seymour. K. Bradford in 1884. May was adopted by her stepfather and changed her name to May Bradford.

After graduating from the Carthage High School, she enrolled in Stanford University in 1898, mainly because at that time there was no tuition. She graduated in 1902, with a bachelor degree in art and mathematics, studying under Bolton Coit Brown, who established the Art Department at Stanford University. Brown headed the department for almost ten years, but was dismissed in a dispute over his use of nude models in the classroom.

May started teaching in Menlo Park and later in Palo Alto while her father founded with his partner Booker & Bradford Mining Engineers shop in Tonopah, Nevada where S.K. Bradford was U.S. Mineral Surveyor. On May 24, 1904, the two men lost everything when the office burned to the ground. They dissolved their partnership but Bradford stayed in Tonopah (Fig. 3.1). He asked his daughter for help. May was a beautiful, slim lady with blue eyes and long dark hair that she wore in a fashionable up-style. She was reluctant to move to the small mining town but finally agreed and arrived in Tonopah on July 21, 1904. She formed with her father a partnership, called "Bradford and Bradford, Surveyors." May was drafting and producing commercial maps of the area. The firm later also hired a draftsman to work under May's supervision.

May Bradford upon her arrival in Tonopah had neither the time nor interest in a social life. She witnessed in Tonopah the worst not only of men but also of women.

© The Author(s), under exclusive license to Springer Nature Switzerland AG 2021
B. Lojek, *William Shockley: The Will to Think*, Springer Biographies,
https://doi.org/10.1007/978-3-030-65958-5_3

Fig. 3.1 Entire population of Tonopah in 1902

Tonopah's population was about two thousand, with twenty-two saloons, dance halls, two newspapers and two daily stages. The harsh Puritan sanctions were not as "practical" as in the more conservative east coast and the city had two brothels with "disgraceful" ladies of the night which were generally tolerated by other women as a "necessary evil." She wrote her mother *"mama I hate men. I know I can never get to the point of marrying."* Her only interest was a horse, "Buck," she obtained as part of compensation for her work (Fig. 3.4). She joined the Tonopah Riding Club and was said to be good with a gun (Fig. 3.5).

In early 1906 May's friend, Marjorie Bowes persuaded May to go to Paris. May followed her studies in Paris under Richard E. Miller at the Academie Julien, the private academy where he and many other American artists studied.

On April 22, 1907, S.K. Bradford's appointment as a U.S. Mineral Surveyor for the district of Nevada was revoked because of a violation of Revised Statutes. Returning to Tonopah by June 1907, May felt she had enough experience to apply for an appointment as Mineral Surveyor. She took an examination and passed. Soon she received a letter from Nevada Surveyor General Matthew Kyle in Reno *"In compliance with your application of the 9th district and the recommendation of the honorable Matthew Kyle of the District of Nevada dated August 7, 1907 you are hereby appointed a Deputy Mineral Surveyor"*.

William Hillman Shockley was born in September 18, 1855 at New Bedford. His father was William Shockley, a whaling captain. His mother Sarah Durfee Hillman was the daughter of shipbuilder Jethro Hillman. She was a descendant of a crew member of the Mayflower, John Alden, who had decided to stay in America (Fig. 3.6).

Fig. 3.2 May Bradford graduation photograph (1902)

Fig. 3.3 May Bradford in her Tonopah "atelier"

Shockley's business was very profitable, earning an average of $500 per month. On January 9, 1859 Sarah took William and sailed to Mauritius to meet her husband. William Hillman learned "*a little French and rode a large tortoise.*" They returned to New Bedford in January 1860.

Later, William Hillman recalled: "*As far as religion went the puritan practices had disappeared from our household. I distinctly remember that one day at dinner*

Fig. 3.4 May Bradford and her horse Buck

Fig. 3.5 Poster for the show performed in Tonopah's dance hall portraying M. Bradford (~1905)
(*…I am an electricity on the gun and I never bat an eye…*)

I asked who made the God? I must have been about ten years old. The answer was unsatisfactory. The question remained fixt in my mind since that. When I was at primary school, I asked the question the teacher. His answer was If I believed as you do, I would commit suicide. I was pulled in a small room as a punishment. I climbed down the water pipe and went swimming. I was rebuked by my father, who did not seem wholly displeased" (sic).

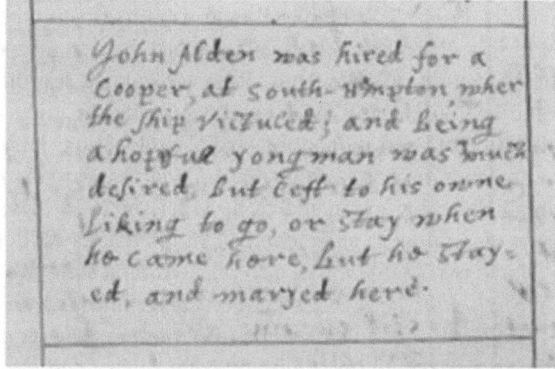

Fig. 3.6 John Alden (from the passenger list of the Mayflower November 9, 1620)

Fig. 3.7 William Hillman Shockley (center) with his brothers George and Walter

All relatives on the Shockley side had somewhat marked individuality; the most marked was Uncle Joseph. Hillman wrote *"He was of extreme simplicity. I often heard him to say that ministers and lawyers were useless members of society."*

When William Hillman was at New Bedford High School, he learned navigation from his father's book and learned how to use logarithms. He excelled in most studies but had a bad habit of arguing with the principal who was *"a man of limited intellectuality, but not a bad disciplinarian as I learned at my cost."* In 1865 William Hillman enrolled in the Bridgewater Academy where he met *"an excellent teacher Mr. Willard"* who tutored Hillman in mathematics. William Hillman's father passed away in 1867 but left enough money to give his son an education. After graduation from high school he decided to go to M.I.T. He went to Boston to take the examination but did not hear anything from the school. In answer to his letter he received a telegram *"admitted without conditions."* William Hillman went to Boston the same day. He rented a room with his classmate L.H. Faulkner, but the second week they had a dispute about the aberration of light. They did not speak for several days although they walked to the school side by side. Faulkner broke the silence and Hillman provided an explanation of aberration; after several explanations Hillman said *"well, I can furnish the explanation, but I cannot furnish your brain to understand it."*

William Hillman Shockley graduated from M.I.T. in 1875 *"and I received my diploma and had no idea how to make money."* In his diaries written several years later he wrote: *"Our class 1875 entered 120 and graduated 24. We learned things in Boston, that are not in the books. Beer saloons and billiards were not infrequent resorts and the taste that I then acquired for liverwurst and Swiss cheese remains. Mathematics was a fairly strong point with me, and I was good in algebra, getting 100% in examination and also 100 in qualitative analysis. When we got to evaluate "Pi" the boys learned it to several places, and I learned it to 50 places."*

After a short seasonal job William Hillman left for Florida as a surveyor, and later to California. From 1880 to 1893 he was employed at the Mount Diablo mine, Candelaria, Nevada, as assayer, surveyor, bookkeeper, and finally as general manager. During this period, he built a dry-crushing silver mill called Shockley & Zabriskie at Candelaria, that held the record for capacity at that time. He operated with his brothers a small gold mine in Grass Valley, California for a few months in 1884 and then with the financial proceeds he followed his biggest interest; for two years he studied languages, literature, art and music in New York and Europe. His brother Walter operated the Candelaria mill until 1897.

While he was in Europe, London based Bewick & Moreing and Co. were advertising for mining engineers to oversee Australia and China, and required that they should be at least 35 years old. William Hillman was 41. He was hired in 1896 as a mining engineer for an expedition to China. The aim of the expedition was to obtain concessions, make loans and establish a mining administration in China. William Hillman took over the Shansi concession in 1898. But because pressure was exerted upon the Chinese Government by the British and due to the increasing unpopularity of the provincial authorities of Shansi, a reward was offered for Shockley's head by the Shansi officials (Fig. 3.8).

During his stay in China Shockley studied the art of calligraphy and assembled a large collection of porcelain and embroidery. After leaving China in 1899, he traveled throughout India and Egypt and went to Vladivostok to explore the Siberian coast. Copper and gold deposits were discovered at Petropavlovsk near the Okhotsk Sea. He surveyed concessions in Australia and Korea during 1901 and 1902. In May to October 1903 he reported on a gold mine in Peru. During the early months of 1905 he investigated a concession in Sudan.

William Hillman Shockley, was one of the first American botanical collectors. Shockley became interested in botany at an early age. Prior to his move to Nevada, he had collected ferns in the limestone sinks around Ocala, Florida. The majority of his plants were found on the wide valley floors and desert mountains within a short radius of Candelaria itself, and around the springs and playas of the immediate vicinity. Nevada was virgin territory for a botanist and many of the plants, even though

Fig. 3.8 William Hillman Shockley with Li Hung Chang (seated), Prime Minister to the Empress of China (1889)

some are now known to be widespread in the Great Basin or southern deserts, were then new to science. By modern standards Shockley's collections were small. He sent one set of specimens to Harvard, where they were studied and named by Asa Gray (*Acamptopappus shockleyi A. Gray*) and Sereno Watson (*Astragalus serenoi shockleyi S. Wats*).

In January 1907 William Hillman Shockley sailed to New York and then took a train trip across country to Nevada. In summer 1907 he met his future wife May Bradford in Nevada for the first time. May wrote to her mother, "*I was amazed to find someone in the middle of Nevada who could talk to me about Italian paintings.*" Hillman was a man of culture with extensive interest in art, music, languages, and literature. He had the style of a gentleman. Hillman was fifty-one years old, while she was twenty-seven. He was smitten with the young lady and May was unconcerned about the age difference. The two dated and then became engaged. They left Tonopah and traveled back to Palo Alto. They married on January 20, 1908 in San Francisco. The Engineering & Mining Journal wrote: "*William H. Shockley of Tonopah, Nevada was married January 20th in San Francisco, California to Miss May Bradford, a young lady having the distinction of being the only woman holding the appointment as a United States Deputy Mineral Surveyor.*"

May and her husband returned to Tonopah and stayed there one more year. William Hillman was not able to draw a profit from his mine and he decided to return back to London. May willingly agreed because she saw it as a new opportunity to study art in Europe. The problem was that the gap between "haves" and "have-nots" was widening, the mining boom started to decline and the Shockley's were often living

only from stock proceeds. May did not know how to cook and she had no interest in managing the household. They kept a cook. When their first son was born, they added a nurse. They did two or three trips to the continent every year.

In April 1913, May and William Hillman sailed back to New York and took a train to San Francisco. They purchased a Victorian vintage house on 959 Waverly Street in Palo Alto. William Hillman lectured on mining at Stanford University and was later elected as chairman of the San Francisco Chapter of A.I.M & M.E. (American Institute of Mining Engineers).

May returned to her passion and started the most productive decade of her artistic life. As an accomplished artist, she exhibited in galleries in San Francisco, Santa Barbara, Los Angeles, and Washington, DC. Her art is now mostly in private collections, but several oil paintings are in the White House (acquired by Mrs. Herbert C. Hoover, a friend of May's) (Figs. 3.10 and 3.11).

May was very loyal to America. During WWI she served as a speaker for the Bureau of Food Administration in Palo Alto, Director of Food Conservation at Santa Clara county, and as Chairman of the Women's Committee Council of National Defense.

In fall of 1922 the Shockley family planned a trip to the East Coast and then to Switzerland. While in New York William Hillman's health worsened and, due to his severe chest pain, May was afraid to leave. The family returned to Palo Alto in April 1923. Later the same year they moved to southern California and bought a house in Hollywood (Fig. 3.12).

May Bradford was a woman of wide culture and diverse interests. She was a fiercely independent woman of high integrity. Her word was binding. She did not tolerate any inappropriate behavior from men in her presence and she did not allow herself to be constrained by the gender roles of her time.

Although William Hillman was a very quiet man, he was a free thinker and held in many ways radical ideas. He was too restless and ambitious to stay in one place

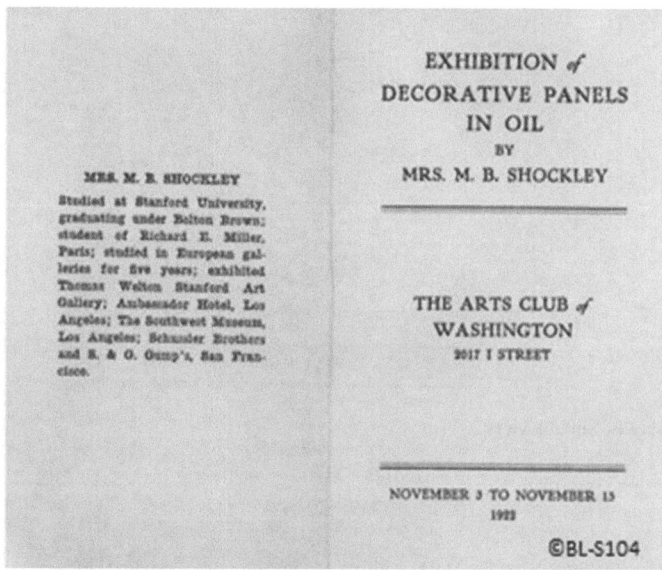

EXHIBITION *of*
DECORATIVE PANELS
IN OIL
BY
MRS. M. B. SHOCKLEY

THE ARTS CLUB *of*
WASHINGTON
2017 I STREET

MRS. M. B. SHOCKLEY
Studied at Stanford University,
graduating under Belton Brown;
student of Richard E. Miller,
Paris; studied in European gal-
leries for five years; exhibited
Thomas Welton Stanford Art
Gallery; Ambassador Hotel, Los
Angeles; The Southwest Museum,
Los Angeles; Schussler Brothers
and S. & G. Gump's, San Fran-
cisco.

NOVEMBER 3 TO NOVEMBER 13
1922

©BL-S104

Fig. 3.10 One-woman show in Washington Art Club (1922)

Fig. 3.11 Commissioned oil paintings of May Bradford

Fig. 3.12 May B. Shockley (1920)

Fig. 3.13 May at home with bust of William Hillman (1976)

for very long. He had the courage of his convictions and he was convinced that the required measures were a necessary step towards a much needed reform of American society. As one of the first members of the Modern Language Association, he directed a campaign for the adoption of simplified spelling.

When the Engineering and Mining Journal published in 1923 a series of columns "On Science and Religion", one of these columns discussed the expulsion of a college professor for teaching evolution. William Hillman sent a letter to the editor with the following paragraph: *"We were hopeful that such persecution would diminish but this hope has not yet been fulfilled. Not only has the ousting of many teachers of modern science continued, but attempt has been made by legislation to stop teaching of science and essentially the teaching of evolution. Attempts to hamper the education of our youth still continue in several states and should be resisted by all who have the slightest interest in freedom of thought."*

Hillman had a strong sense of economic and social justice of miners, he defended them frequently. The episode "The American Institute of Mining Engineers as Censor—A Protest" illustrates best William Hillman's attitude. A member of the managing committee of the 1915 International Engineering congress, H.F. Bain, asked William Hillman to present a paper on the economics and sociology of mining. Hillman in agreeing to write a paper informed Bain how he expected to treat the subject. The paper titled *"The Economic and Social Influence of Mining"* was passed by the editors and accepted by the publication committee of the Congress and it was printed. Hillman presented his paper which contained a table showing average yearly earnings of native-born miners as $732.00 and foreign-born as $447.00, with an estimate of The United States Bureau of Labor and Statistics showing that in order to support a family decently the income should be not less than $750.00/year. Operators of the mines objected to Hillman's presentation and demanded the withdrawal of the table from his paper. On June 23, 1915 the directors of Mining Engineers voted that unless William H. Shockley changed the paper to meet the views of the mine operators, it should not be published.

Hillman wrote a sharp two pages protest against censoring his paper which was published and stated *"to suppress these well-known statistics seems to me not only futile but stupid"* and continued *"My own opinion is that in denying me a privilege of quoting official reports you lay yourself open to the reproach of taking a partisan*

©BL-S128

Fig. 3.14 William Hillman Shockley (few months before his death in 1925)

view of a disputed question and violating the ancient legal maxim: Audiatur et Altera Pars." The published protest triggered responses from many members of A.I.M.E. who protested against the action of the Institute in issuing a decree against one of its own members. One of the responses reads *"as the matter stands, Mr. Shockley's note reads very much like a famous recantation of Galileo and its closing paragraph recalls his E pur si muove".*

After moving to Hollywood, William Hillman's health continued to decline. He had to stop working in early 1925. From April, the doctor visited him daily but his condition did not improve (Fig. 3.14). He passed away on the evening of May 26, 1925 with May and his son at his side. He left to his wife and son stock and bonds worth over $70,000. In a 1970 interview for the Hoover Presidential Library, May stated, *"I never met a man who was worthy to unfasten the sandal of William."* May passed away of natural causes on March 7, 1977, at the age of 98. She was mentally fit until her last day. Her son inherited a modest trust fund and land in Hawaii.

William Hillman Shockley was a kind of renaissance man fluent in six languages and enlightened in the arts, music, literature, and science. Hillman was well read and kept a large library; with his wife they discussed art and literature. He loved his wife dearly, and May and William Hillman were happily married. May frequently asked Hillman to re-tell his stories from when he travelled the world.

Chapter 4
The Son Billy

*"To be a little better than others does not require a little work—it
requires a lot more work."*

Ignacy Jan Paderewski (1860–1941)

The Shockley's were living in London when William was born. His father wrote in
his diary on February 13, 1910:

```
At just 10 o'clock I heared the cry of the baby, a strong penetrating
lusty cry. And a long time latter the nurse came out, and said "It is a
little boy."     About 10:30 ----------I saw the baby in his cotall all
wrapped up. I was surprised to find how plump and stout he looked, and
how much of humanness was in his face..
    The baby weighs eight pounds, and has a small amount of dark brown
hair,and seems to have dark blue eyes, but has not opened them much yet.
```

Two days later an additional note says: *"Judging by what little he has shown it is
probable that he will be a vigorous child"* (Figs. 4.1 and 4.2).

The family returned to California in 1913. Shockley did not enter elementary
school at the usual age, however, because, as he frequently said: *"My parents had the
idea that the general educational process was not as good as would be done at home."*
May and William Hillman loved their son but they had their own ideas about how the
child should be brought up. They were very proud of their son and they loved him
dearly. Father wrote in his diary: Billy *"is going ahead mentally very fast and is very
well."* But they kept him out of school until he was eight years old. His mother taught
him mathematics, and his father, who was lecturing mining at Stanford University at
that time, discussed with him topics such as geography, religion, space, or literature.
Both parents encouraged his scientific interests and often they treated Billy as an
adult. Father Hillman took parenting very seriously: he enrolled in a correspondence
parenting class offered by A.H. Putnam at the University of Chicago. When asked
who was punishing the boy, Hillman answered *"this is mama's role, she has slender
bamboo sprout and sometimes speaks to Billy forcibly."*

Social interaction helps young children to start to develop their sense of self, and
also to start to learn what others expect from them. Billy had very little contact with
other children, and had not developed any social skills or the ability to have empathy.

© The Author(s), under exclusive license to Springer Nature Switzerland AG 2021
B. Lojek, *William Shockley: The Will to Think*, Springer Biographies,
https://doi.org/10.1007/978-3-030-65958-5_4

Fig. 4.1 Billy in London 1912

Fig. 4.2 William Shockley, Mom and Dad

His father always emphasized *"you always tell the truth"* or *"lies are always more complicated than truth"* and Billy always did tell the truth. He always said "it is what it is." In his adult life Billy never understood why in certain situations telling the truth causes problems.

Bill's interest in physics developed early, inspired in part by Professor Parley A. Ross, a Stanford physicist and the Shockleys' neighbor. Ross exerted an especially important influence in stimulating Bill's interest in science. Billy was a frequent visitor at the Ross home, playing with the professor's two daughters, and became a substitute son.

Fig. 4.3 William Bradford Shockley (September 1921)

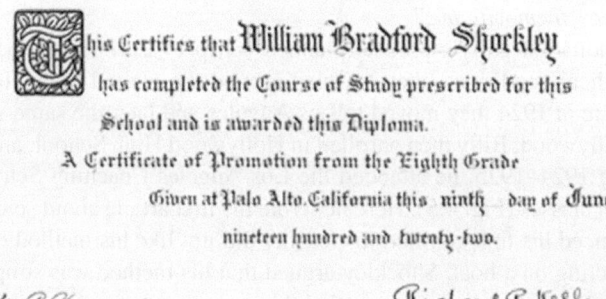

Fig. 4.4 Billy's Palo Alto Military diploma (June 1922)

May wanted to know how they were doing with Billy's home schooling. She requested Stanford Revision of the Binet-Simon Test with Lewis Terman as examiner. On January 1919, May scored 161, and nine year old Billy 124.

Billy spent two years at the Palo Alto Military Academy. The Academy was a 1st through 9th grade all-male boarding and non-boarding military school where superior scholarship, character, and leadership were emphasized. Billy's final evaluation reads: *"curious, punctual, good student, understands the concepts rather than just memorizing them"* (Figs. 4.3 and 4.4).

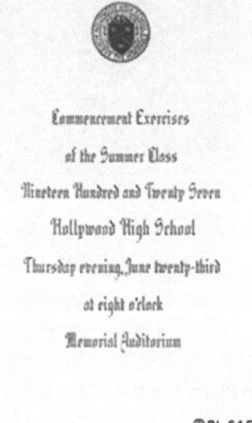

Fig. 4.5 Billy's High School graduation photo (1927)

Through Billy's formative years, his own father was a strong and constant influence. Shockley later stated: *"He encouraged me into scientific studies and would always discuss them with me."*

After graduation Billy's parents planned a trip to Europe, but because of his father's medical conditions, they traveled around the United States for more than one year. Late in 1924 they moved to Los Angeles and later the same year bought a house in Hollywood. Billy then enrolled in Hollywood High School, and at the same time, during 1924–1925, he attended the Los Angeles Coaching School where he took physics classes (Fig. 4.5). Here he wrote his first article about science teaching and experienced his first conflict: his teacher did not like his method of calculating the forces acting on a boat. Shockley argued that his method was simpler and gave the correct answer, but they never settled this argument.

In early 1925 Billy's father passed away just before Billy gained admission to the southern branch of the University of California. Billy had excelled in science and mathematics at Hollywood High School. He had completed school at the age of seventeen and he had made his commitment to a career in physics. Before attending the California Institute of Technology (Caltech) he spent a year at the University of California at Los Angeles (UCLA).

Billy transferred to the California Institute of Technology in 1928; by then he was a brilliant student who distinguished himself in another way. To maintain an athletic body, he followed a laborious exercise regime. Billy divided his time between studying and the gym. Al G. Treloar, a Harvard-educated bodybuilder who was director at the Los Angeles Athletic Club, provided Billy with free access to the Club in exchange for the use of Billy's photographs in Treloar's advertisements.

CalTech was, from early 1920, a great educational organization with science giants such as Robert A. Milikan, William V. Houston, Richard C. Tolman, and

Linus Pauling. Russian born Boris Podolsky, who held a National Research Council fellowship at Caltech and had the uneasy task of providing tutorial supervision in quantum mechanics to the promising undergraduate student William Shockley, said *"we can never finish on time because Shockley continually asked questions one after the other."*

When Shockley returned to Caltech as visiting professor in 1954, he said *"Dr. Tolman was a philosopher with interest in other subjects, Linus Pauling was stimulating in any subjects. But the best of all was Dr. Houston. He had an unusual capacity to see what was in the student's mind. In the way some student would ask some question which was quite confusing, Houston would listen to it, and then he would formulate the question so it was clear how the student should have done it, or what was mixing him up, and then he would answer it. So, he had a remarkable capacity to grasp the confusion in the student's mind and see just what was needed to straighten it out."*

Shockley earned a bachelor's degree in physics from Caltech in 1932 and was offered a teaching fellowship at the Massachusetts Institute of Technology. While working on his graduate degree, Shockley did Physics class teaching. His salary was $80.00/semester but living expenses, room rent, and parking for his 1930 DeSoto car were about $150/semester. The balance of the required income was left to May. Shockley was teaching three undergraduate classes, each 3 h/week, and took in an average of 16 credits per semester. For the first time he had no time for the gym, he worked around the clock. Inspired by CalTech professors, Shockley updated the curriculum and designed several demonstration aids for his courses. This work triggered his life-long interest in mechanical engineering and resulted in the very first of Shockley's publications.[1,2] One of the memorable experiences which Shockley often described was his encounter with one of his students who had *"much grasp of mathematical relevance."* His name was Richard Feynman. Dealing with his students triggered another of Shockley's interest: how to deal with the education of students. He was convinced that the perfect lecture was not necessarily the best method of teaching. Shockley shared an office with another of John Slater's students, Robert D. Richtmyer, who was also studying physics although he eventually turned to mathematics. But his best friend was James Fisk, who had the same taste for misbehaving as Shockley. They glued a professor's triangle to the desk, they re-wired elevator buttons, so the elevator stopped at a different floor, and they attached a harmonica on the grille of Prof. E. Condon's car so that it would start blasting out when the car exceeded a certain speed.

The title of Shockley's doctoral dissertation was *"Calculations of Wave Functions for Electrons in Sodium Chloride Crystals"* (Fig. 4.6). Shockley later said that it was this research in solid-state physics which *"led into my subsequent activities in the transistor field."* Shockley's thesis advisor was John Slater, a kind of father

---

[1]W. Shockley, "Application of an Electrical Timing Device to Certain Mechanics Experiments," Am. Journal of Physics Vol. 4 (1936), pp. 76–80.

[2]R.P. Johnson, W. Shockley "An electron microscope for filaments: Emission and Absorption by Tungsten Single Crystal," Phys. Rev. Vol. 49 (1936), pp. 436–440.

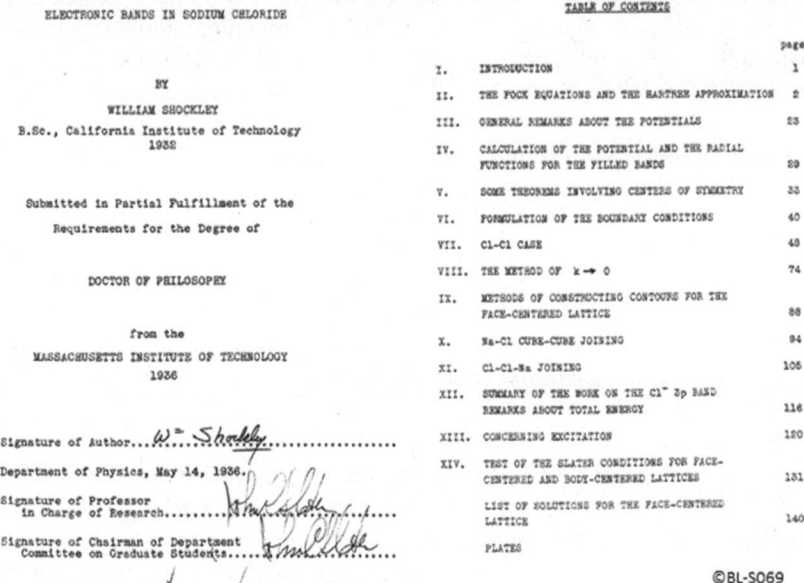

Fig. 4.6 W. Shockley's Ph.D. thesis 1936

of American solid-sate physics, who was the head of the MIT Physics Department. John Slater, although one of the top ranked lecturers with a capacity to explain things simply and straightforwardly, was a very private person who kept a certain distance from his students. His lectures were humorless: if he was in good mood, he entertained his students simply by counting, slowly and solemnly, up to forty in Danish. John Slater was somewhat critical of refugee physicists emigrating from Nazi occupied countries. He insisted that even before the arrival of the Bethe, Bloch, Brillouin, and others, American physics departments provided better education in quantum theory than their European counterparts. Shockley did not share the opinion of his thesis advisor. He studied publications by Bethe, Bloch, Hartree and Fock. His bible was the 1934 translation of the Russian textbook "Wave Mechanics: Advanced General Theory" by Yakov. A. Frenkel.

Gradually, unsatisfied with Slater, Shockley sought advice elsewhere and found his best mentor—Philip M. Morse. Morse was only a few years older than Fisk and Shockley. He was unconventional and flamboyant, a tall, handsome man intent on good living. He loved good food, women, and music. He formulated a theory that creativity and sexual energies were connected. A line he frequently used with his lovers was *"sexuality is the greatest gift we have been given."*

Using the modified Hartree-Fock equation Shockley calculated the highest filled energy band of crystalline sodium chloride. He did all calculations using a mechanical calculator and the results were "hand plotted," as shown in Fig. 4.7.

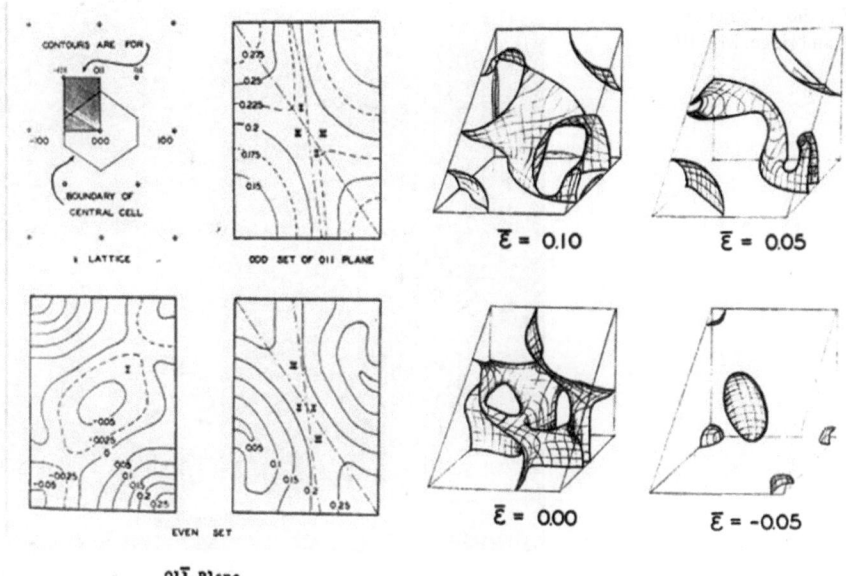

Fig. 4.7 Hand drawing diagrams from W. Shockley's MIT thesis (1936)

He also suggested a better approximation for a plane wave solution inside the Brillouin zone, but justified not actually performing these calculations as follows: *"It would be interesting to test such improvements by the method of this section, and the writer has been prevented from doing so only by the discrepancy between the Earth's rate of motion about the Sun and the writer's rate of doing work."*

Shockley took Morse's nuclear physics courses and they published two papers[3] but Shockley's primary interests were in theoretical calculations for solids. Shockley's thesis was the very first attempt to draw a realistic picture of complex energy bands for a real crystal, or what is now called "electronic structure calculations." Shockley had also shown in his thesis that the Wigner-Seitz cellular method (1933) leads to large errors for an empty lattice for which the value of the energy as a function of wave number is known.

His Ph.D. thesis was for the first time signed W = Shockley.[4] The idea originated from R. Richtmyer after he and Shockley discussed Group Theory (Chap. VII in Shockley thesis).

---

[3] J.B. Fisk, L.I. Schiff, W. Shockley, *"On the binding of neutrons and protons I"*, Physical Review 50 (1936), pp. 1090–1091.

P.M. Morse, J.B. Fisk JB, L.I. Schiff, *"Collision of neutron and proton II"*, Physical Review 51 (1937), pp. 706–710.

[4] John Bardeen later used the signature Bardeen $\leq$ X in communication with Shockley.

Fig. 4.8 Mother and son
(Cambridge, MA 1936)

Shockley defended his thesis on May 14, 1936 and graduated on June 26, 1936 with May, Jean, and he daughter Allison in the audience. The 1936 Class graduating party was held that night with Morse present, and no one invited Slater.

Chapter 5
Jean and Emmy

*"I don't want to be married just to be married. I can't think of
anything lonelier than spending the rest of my life with someone
I can't talk to, or worse, someone I can't be silent with."*

Mary Ann Shaffer
The Guernsey Literary Society

Time is required for two people to get to know one another. While you can lust
after what you do not know, you cannot love what you do not know. Knowledge is
a prerequisite for love, but this is something that no one can understand when they
are young.

Bill met Jean Alberta Bailey for the first time when they were both students at
the southern branch of the University of California (now UCLA). Jean was born on
June 13, 1908 in Cedar Rapids, Iowa. Jean grew up in a very religious family. Her
father Bert H. Bailey, the son of a Presbyterian minister, studied at Rush Medical
College in Chicago and anthropology at the University of Iowa. After graduation, he
accepted a position at the Zoology Department of Coe College in Cedar Rapids and
lectured on ornithology. He has been described as a scientist in a constant attitude of
worship. Jean's mother, Anna Condit, was a housewife. When Jean's father, Bert H.
Bailey, passed away in 1917, her mother Anna moved to Los Angeles to be close to
her brother.

At that time, when Bill and Jean met, Jean was dating someone else. They met
again at the spring of 1933 when Bill returned for the summer to Los Angeles. By July
Jean was pregnant. They were married in August 27, 1933. The couple's honeymoon
days at Two Harbors in Catalina Island were amazing. They were infatuated with
one another. Their lust, fueled by idealization, prevented each from seeing the real
person or any flaws. May, when she met Jean for the first time, saw no reason to
change her view of the subject of what she considered her son's disastrous marriage.

Bill and Jean kept the pregnancy secret until Christmas. When May learned about
the pregnancy, she was very angry and disappointed. She saw the differences between
Jean and Bill, but Jean and Bill did not want to see faults in each other and they did
not count the cost they would pay later. Jean and Bill's daughter Allison was born
on March 25, 1934.

While Bill continued his studies at MIT, newlyweds Bill and Jean moved to
Cambridge and rented an apartment. They struggled financially, after paying the rent
and nursery school for Allison there was not much left. May was sending monthly

© The Author(s), under exclusive license to Springer Nature Switzerland AG 2021
B. Lojek, *William Shockley: The Will to Think*, Springer Biographies,
https://doi.org/10.1007/978-3-030-65958-5_5

checks and parcels with food and clothing over the next two years to support Bill until he completed his studies.

Shockley was hired by Bell Laboratories in 1936. His salary was $325.00/month, which was a lavish salary at that time. They rented an apartment in Brooklyn but after paying rent and nursery school for Allison, there was still not much left. Jean found a part time job at the YWCA, an organization that issued a call to pray and act together in solidarity with members and partners around the world. Jean read the bible daily.

Yet, the years in Brooklyn were the best years of Jean and Bill's marriage. Shockley returned to his exercising habits, they spent weekends climbing and hiking in Upstate New York. After Bell Laboratories moved from Brooklyn to Murray Hill, the Shockley's moved to Gillette, NJ in the summer of 1939 and rented a house there. Bill bought a used 1934 car LaSalle Series 50. Jean was not good in managing the household and Bill, after their frugal years at MIT, did not like her obsession with shopping and buying shoes. Jean started taking a course in educational psychology at Montclair State Teachers College, in Upper Montclair, NJ, hoping to find a job and subsidize her shopping needs, but she did not finish it.

When the landlord raised the rent, the Shockley's rented a smaller house. Shockley was working long hours and had to travel frequently, but he was a good father to his daughter and spent as much time as he could with her. Bill read books to Allison which Jean considered inappropriate for a six year old girl. He did not like Jean trying to expose Allison to religion. He asked Jean to keep religion out of the household. Despite the differences between Bill and Jean, in August 23, 1942 Shockley's first son William Alden Shockley was born.

The relationship between Bill and Jean was in many ways never a close one. They were in ardor in the very short courtship stage of dating, with little concern for the other person's attitudes, opinions, or concerns. Jean admitted later that she and her husband never really got to know one another and that Bill took a very small part in bringing up his children. Bill, like all eminent creators, possessed not only a superior intelligence but also a considerable ego and other traits of personal fortitude and self-discipline. Such attributes produced the kinds of symptoms that make relationships difficult. Bill was very seldom able to confide any of his thoughts to Jean. They drifted apart. Bill and Jean had different kinds of failings, for very different reasons. Bill prioritized work over marriage, family, and togetherness. Jean demanded too many things that were basically foolish; she did not understand that creative people usually spend more time on their own because it allows them to focus on thinking. Bill was highly disciplined and rarely wasted time; for him life was a serious business and individuals should be purpose driven.

In spring 1942 Bell Laboratories put Shockley on leave. He joined the Anti–Submarine Warfare Operations Group (ASWORG) organized by Phil Morse. During the war, Shockley often traveled on business assignments and he did not see his family sometimes for several months.

When World War II ended, M. Kelly, the head of research at Bell Laboratories, sent a request to the Pentagon to release Shockley. When Shockley returned to Bell Laboratories in September 1945, his salary was increased to $950.00 and he resumed the pre-war physics work in Bell Laboratories Physics Department.

Bell Laboratories researchers frequently socialized, and there was a kind of camaraderie, partying and spending time together. Jean could not keep up in discussions with some other wives and her religious views were quite often embarrassing to Bill. Jean considered Bell Laboratories people snobbish, while they looked upon her and characterized her as a "nice housewife." Bill realized that he had married the wrong woman. Yet, undeterred by mounting marriage problems, a second son Richard was born on September 6, 1947. The same month they bought a new house close to the Madison train station. For the first time, Shockley could have an office at home.

Jean was not a good student and she never pursued any higher intellectual goal. Her only interest was bird-watching. She expected someone else to make decisions for her and her world reduced to reading Reader's Digest. Jean assumed that children were a gift from God and that God would raise her children. She never disciplined the children. Bill, on the contrary, believed that discipline is not only good for children, but is necessary for their happiness and self-discipline. When a marriage is not good, it not only takes a toll on the children but it can feel like an almost bottomless source of loneliness. Bill was going through a very difficult time, and his only confidant at this time was May. He wrote in a letter dated January 14, 1951 "*I know it is going to be a catastrophe.*" Bill had been living in his own inner world where he shut himself away from Jean's unreasonable demands.

In April 1951, William Shockley became a member of the National Academy of Sciences, one of the youngest scientists ever to attain this honor. In 1953 the American Physical Society awarded Shockley its first Oliver E. Buckley Prize for advancements in solid-state and condensed-matter physics. And in 1954 he received the National Academy's prestigious Comstock Prize, awarded every five years for major advances in electricity and magnetism. But these achievements and honors came at the cost of mounting disaffection from his family. Although close to his daughter Allison, Shockley grew distant from his two sons William and Richard, who were much younger than their sister.

There was a brief marriage reconciliation when Jean was diagnosed with uterine cancer in 1952. Shockley, equipped with knowledge of nuclear physics, and through contacts with members of the National Academy of Sciences, shared Jean's medical records with the leading experts on uterine cancer in Washington, Boston, and New York. To decide whether radiation or surgery would be the best under the given circumstances, Bill studied all that was in his reach. He fired Jean's doctor and hired a new group of specialists in New York; on June 1953 Jean had the last radiation treatment. During that year Bill took care of the three children and visited Jean in hospital regularly. Jean fully recovered, did not experience further uterine cancer, and lived another twenty-four years. She passed away in 1977. Very likely, Bill's

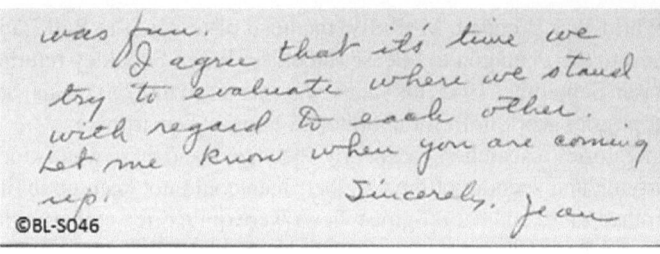

Fig. 5.1 Jean's letter to Bill dated February 5, 1955

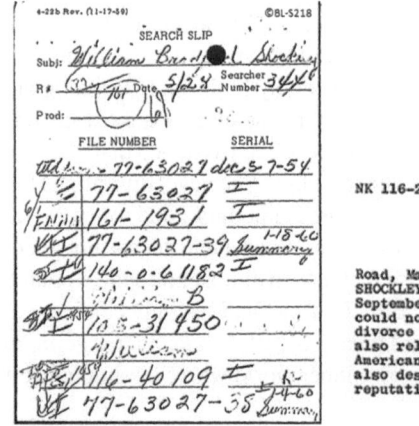

NK 116-2461

MISCELLANEOUS

On March 21, 1962, Mrs. JEAN SHOCKLEY, 22 Academy Road, Madison, New Jersey, advised that she was married to SHOCKLEY in 1933 and obtained a divorce from him in September of 1955 at Reno, Nevada. She stated that she could not recall the grounds on which she obtained the divorce and that it was an entirely personal matter. She also related that she considers SHOCKLEY to be a loyal American citizen of high type character. JEAN SHOCKLEY also described her former husband as a person of excellent reputation and associates.

©BL-S219

Fig. 5.2 Jean's file in William Shockley's FBI Record[1]

radical approach saved Jean's life. He demonstrated once again that he could grasp the root cause of any problem very quickly and find the remedy.

Neither Bill nor Jean were unfaithful, but they both realized that they had married the wrong person. They decided on a divorce (Fig. 5.1). Bill and Jean agreed on an amicable divorce settlement in the spring of 1955 and they divorced on August 30, 1955. They were able to maintain a decent relationship for the rest of their lives (Fig. 5.2).

Allison graduated from Radcliffe; William dropped out of Lehigh University. The youngest son's custody was transferred to his father because Richard ran into disciplinary problems. He moved to Palo Alto and studied physics at Stanford.

When Bardeen, Brattain, and Shockley were awarded Nobel Prizes, Jean telegraphed: "*Just heard tremendous news re Nobel prize as Newark News phoned for story to amplify as dispatch phoning Allison at once congratulations don't have tell*

---

[1] W. Shockley was a member of *The American-Soviet Science Society*. The House Committee on Un-American Activities required that the FBI establish a record of all members of the society.

you how I feel may the laurels be always green and beautiful and all happiness to you and Emmy." (Western Union telegram, November 1, 1956, 7:40AM).

Emily "Emmy" Ina Lanning was born on December 15, 1913 in one of America's coldest winters, in the small town of Cazenovia, New York. Her father managed the family owned and operated Cazenovia Lumber & Oil company in Cazenovia, located on the shores of Cazenovia Lake in Central New York. Her mother ran a small grocery store. After finishing high school Emmy worked for a while with her father and then enrolled at St. Lawrence College, Kingston, Ontario. She practiced nursing training at Bellevue Hospital where Columbia University's College of Physicians and Surgeons had assigned faculty and medical students. In Bellevue, Emmy developed a strong interest in psychiatric nursing and she enrolled at the Catholic University in Washington, DC for her Bachelor and Master's degree. After graduation she was offered a job at Chestnut Lodge, in Rockville, Maryland, a psychiatric sanatorium well known for the care of nervous and mental diseases. The Lodge was a private mental hospital with an unusually high ratio of nurses to patients, serving approximately seventy patients at a time. At Chestnut Lodge a Russian-Jewish immigrant Morris S. Schwartz, a research sociologist, combined efforts with Emmy. As a psychiatric nurse and director of a nursing program, she developed a program which became the model for other psychiatric institutions and which the subject of the book *"The nurse and the mental patient."* Emmy and Morris published this in 1956 [2] and when it was published, Emmy was listed as coauthor Emmy Lanning-Shockley.

Joan Ascher, who was Assistant Director of Nursing at Chestnut Lodge and a rock-climbing friend of Shockley's, and her fiancé Phillipe V. Cardon, invited 41 year old Emmy and 44 year old Bill for a dinner party on November 18, 1954. Emmy was a warm and exuberant woman who could sweep a man off his feet. She was well accustomed to powerful intellectual men. During the evening, Shockley shared with Emmy the draft of his presentation about the effectiveness of research laboratories, to be given the next day at the Operations Research Society of America Meeting in Washington, D.C. Emmy was familiar with methods of statistics and when Shockley asked her for feedback, Emmy disagreed with one of the statements and she spoke up. Then they talked for hours. In the middle of the night Emmy offered Bill a ride. When driving, she asked him "are you married?" Shockley showed her a wedding ring which he wore and answered truthfully. She dropped him at the University Club, where he was staying. Shockley asked Emmy if she would like to hear his presentation with changes based on her input. Emmy agreed.

When Shockley returned to Pasadena, they started exchanging letters almost daily. Shockley knew how to court a woman. Shockley could recite the poetry of T.S. Elliot with different speaking parts for half an hour, or a piece of Mark Twain literature. He sent Emmy flowers regularly. He fell in love with Emmy. Shockley always had something special, something appealing for Emmy.

---

[2]Morris S. Schwartz, Emmy Lanning-Shockley, R.N. Russell Sage Foundation, New York, 1956.

Fig. 5.3 Bill and Emmy in London (1957)

Shockley accepted a consulting job for the Pentagon, and rented a State House Apartment in Dupont Circle, Washington D.C. because he wanted to be close to Emmy. The problem was that Emmy had accepted a position at Ohio State University in Columbus and left Washington on December 1, 1954. They continued correspondence and seeing each other on weekends except when Shockley was in Madison. On January 17, 1955, Emmy wrote in a letter to Bill *"I want you so much - so very, very much."*

Unhappy with his administrative job at the Pentagon, Shockley considered a university position. In March 1955, Jack Morton and Shockley discussed over drinks the future of the transistor. Jack shared with Bill the advances in the silicon float-zone refining methods and progress in diffusion achieved by Fuller and Morris Tanenbaum and suggested to Shockley to start his own company. At this time, Shockley had committed to lectures in London and the Pentagon-related work in France (Fig. 5.3).

When he returned to Washington in May 1955, he immediately traveled to Columbus to meet Emmy and presented her with his plan to separate from Bell Laboratories and start his own company. In the middle of the discussion he told Emmy that he and Jean had agreed on a divorce. He separated form Bell Telephone Laboratories effectively August 31, 1955. At the age 45, deeply in love with Emmy, William Shockley sold his beloved Jaguar XK120,[3] finalized the divorce, and left his job with a plan to start his own company.

---

[3]The legendary XK series, the Jaguar XK120 was introduced in 1948, and exceptional styling and inspiring performance made the model an instant classic. With its name alluding to its top speed of 120 mph, the XK120 was one of the fastest cars of its time. When the first XK120s arrived in North America, their beguiling design attracted such Hollywood-based customers as Clark Gable, Mamie Van Doren, and Jayne Mansfield. Shockley bought his car in London in 1953.

Fig. 5.4 Bill and Emmy in Germany (1956)

Emmy and Bill were married in a civil ceremony on Thanksgiving Day 1955. Emmy's contract in Columbus expired in March 1956 and she moved to Palo Alto.

When asked why he married her, Shockley answered *"because she understands people better that anyone I know."* At a private ceremony after Bill passed away, Emmy told Arthur Jensen *"Bill proved that I can fall in love again, again and again."* Those who knew them often saw their relationship as "us against the world." They felt so linked together that they were ready and willing to take on any feat in life, so long as they had their soulmate by their side.

Emmy was the best thing that ever happened to William Shockley. Emmy was not only Bill's wife and lover, she was his cohort, companion, trusted friend, assistant, and unwavering supporter. She was totally absorbed in him and his work, even though she had her own very strong personality, quite different from his. When you saw Emmy and Bill together you knew instantly that they were not separable. They did not talk about love, they showed love to each other, they lived in love and their love was genuine. When they traveled, they strolled together, hand in hand. Bill and Emmy felt comfortable together in silence, they could communicate intuitively, without talking, each feeling the other's presence. Contrary to Jean, Emmy learned to be efficient at managing Bill's time. In addition, Emmy possessed skills which Bill did not have. She had creativity, sensitivity, humor, irony and a gift for graceful solutions to the kind of *faux pas* that direct-shooting Bill could so easily cause. Shockley was completely dependent on Emmy. It was Emmy's support and unfailing loyalty that kept him sane for all those years, and for that, he loved her.

Although Emmy would live for another eighteen years after Bill passed away, she would never find another, and would not stray. She changed very little in the house. Their two gardening hats remained in the entrance hall as always.

Chapter 6
Bell Telephone Laboratories

> *"Most people say that it is the intellect which makes a great scientist. They are wrong: it is character."*
>
> Albert Einstein.

After defending his thesis on May 14, 1936 Shockley was offered a position at Yale University. At that time, Mervin J. Kelly, the new director of research at Bell Labs, began recruiting solid-state physicists. He called Shockley to set up an interview. Shockley and Fisk were recommended to Kelly by Phil Morse. Kelly promised work in C. J. Davisson's group, so Shockley declined Yale's offer and accepted Kelly's. Shockley joined Bell Laboratories located at the Western Electric building at 463 West Street, New York on Monday, September 14, 1936, and reported to Clinton J. Davisson. Davisson led scientific teams doing research on the high-volume production of vacuum tubes (the official title of the group was "Vacuum Tubes and Transmission Instruments").

Shockley's effect on vacuum-tube research was galvanizing in his work with others. Shockley was of the next generation of physicists who were particularly adept at applying quantum mechanics to specific problems. He often discussed and explained it to the older physicists, for whom quantum mechanics was a completely new thing. Shockley organized the "Journal Club" of informal discussions and one night each week they discussed the basics of quantum mechanics and solid-statephysics with Frenkel's and later with Mott and Jones' book[1] as the main textbooks. There was an auditorium on the top floor where these different disciplines met to educate each other, with tea and cookies provided by the management. It was held after work hours because it was entirely their own show, informal, lively, and tremendously useful. Einstein and every other eminent European scientist who went to New York came to visit or give talks. The only rule was clarity—no footnotes, no jargon. Later Brattain recalled *"I always enjoyed being a member of this group. I was impressed by the fact that not all members of this group were Ph.D's. But they*

---

[1] V. Y. Frenkel, *"Wave Mechanics. Advanced General Theory,"* Clarendon Press, Oxford, 1934.
N. F. Mott, H. Jones, *"Theory of the properties of metals and alloys"*, 1st edn., Oxford University Press, 1936.

© The Author(s), under exclusive license to Springer Nature Switzerland AG 2021   43
B. Lojek, *William Shockley: The Will to Think*, Springer Biographies,
https://doi.org/10.1007/978-3-030-65958-5_6

Fig. 6.1   John R. Pierce

all carried weight in the discussions." The group studied not only physics, but also the Russian language.

The Caltech electrical engineer, John R. Pierce, also a new employee, who had joined the tube department at the same time, described the collaboration with Shockley[2]: *"Certainly, Shockley had a lot to do with the construction of electron multipliers. He came to the rescue, and introduced me to Liouville's theorem, which says that the density of particles in the coordinates of phase space is constant. If we regard the motion of electrons as governed by the electric (and magnetic) fields in which they move, and disregard interactions of one electron on others, this applies to electrons in vacuum tubes, and the coordinates are the three position coordinates and the three velocity coordinates. The derivation of Liouville's theorem that Shockley gave me used Hamilton's equations, about which I knew nothing, but I found that the theorem could be derived easily from Newton's laws of motion, and I suppose that's how Liouville derived it. It was because of him that a practical means for designing electron multipliers was worked out. I had come to have a high regard for Bill Shockley, who was good and helpful to me, and was often good fun as well"*.

On October 13, 1937, Shockley and Pierce issued a joint memorandum *"Maximum Attainable Current Densities and Deflection Type Tubes"*, in which they described how electron multipliers work, and submitted a patent application.[3]

Another member of the Davisson group, A.L. Samuel, asked Shockley how he did all the calculations in his Ph.D. Thesis. When he learned that Shockley used a

---

[2]John R. Pierce, Interview by H. Lyle, April 16, 1979, Caltech Archives.

[3]J. R. Pierce, W. Shockley, Electron Multiplier, US Patent 2,245,606 (November 26, 1937).
W. Shockley, J. R. Pierce, *"A theory of noise for electron multipliers,"* Proc. IRE, Vol. 26 (1938), pp. 321–332.

Fig. 6.2 Bell Telephone Laboratories, Murray Hill, NJ

mechanical calculator manufactured by the Monroe Calculating Machine Company,[4] Shockley's first assignment in the Vacuum Tube group was calculation of the space charge between parallel plane plates.[5]

At the end of 1938 the Western Electric laboratory moved to Murray Hill, New Jersey. Bell Telephone Laboratories was formed as a subsidiary corporation (Fig. 6.2). The corporate organization recognized that it was necessary to put men trained in the scientific method in charge of industrial research. Dr. Jewett of the Bell System was the first scientist engaged in managing creative technology independently of the corporate manufacturing section of the Western Electric Company. When Shockley joined Bell Telephone Laboratories, the organization had 2200 scientists, engineers, and additional supporting staff.

---

[4]W. Shockley authored several application notes for the Monroe Calculating Machine Company while at MIT.

[5]C. E. Fay, A. L. Samuel, W. Shockley, "Solution of space charge region between parallel plates", BSTJ, Vol. 17 (1939), pp. 49–79.

Fig. 6.3 Dr. Mervin Kelly,
Research director of Bell
Laboratories (1937)

In January 1940, Kelly created a new "Physical Research Department" headed
by Shockley and reporting to Harvey Fletcher. The other members of the group
were metallurgist Foster Nix and physicist Dean Wooldridge. Kelly's vision was that
the group would do the basic research in solid-state physics. Nix had previously
been a member of the Davisson group and studied in Germany, working on metal
physics, while Dean Wooldridge, a Caltech physics graduate, had worked on electron
emissions.

Foster Nix wrote: *"Shockley and I did work on self-diffusion of copper together,
and we did the work on the order disorder transformation in Reviews of Modern
Physics which I invited him to participate in. He handled most of the theory and I
handled all the experimental work. But I was the senior one. I was writing it before
Shockley came into the picture, and I realized he could do a much better job and
it would be a much better paper, by having someone who was more versatile and
had the machinery of theoretical physics—than I was. And it was a better paper by
virtue, no question about that"*.

Mervin J. Kelly (Fig. 6.3), a fanatic tulip gardener, thought he could judge a man
after a few moments of conversation, and he judged people and programs by real
accomplishments. Kelly had a temper that frightened many, he would turn red and
let everybody know where he stood. He did not tolerate "yes" men. No one resisted
Kelly. When he considered someone was blocking his way, that person was severely
mistreated. The success of Mervin Kelly was that he only had one requirement for his
employees: perfection. Kelly's philosophy was that the researcher's freedom would
lead faster to the new practical application. Kelly believed that the best science and

technology resulted from bringing together and nurturing the best minds. He worked diligently to acquire an extraordinary collection of scientists and then provided them with the support to create the breakthroughs that defined the industry. He understood the needs and complexities of the exceptional mind. *"It's exceedingly unlikely to find multiple talents in a single person,"* he wrote in a report to management, *"but it is in the mind of a single person that creative ideas and concepts are born."* Kelly had an exceptional ability to manage what are known in the current terminology of Human Resources Departments as "difficult people". Kelly understood that the type and quality of "difficult people" selected for research, the environment the company provided, and freedom equivalent to those doing research in academia, are the most important factors in the effectiveness of research.

Shockley described Kelly: *"I think developing [an] optimum atmosphere depends upon an understanding of [the] spirit of research people. This understanding rarely develops in a man who achieves high managerial statute without personal technical experience."*

This did not mean wasting company resources. Bell Telephone Laboratories was efficiently organized. Kelly had a secretary and clerk, while everyone else shared the Drafting and Typing Department. The technicians and facilities were shared by all groups. There was nothing like what Shockley later characterized as a *"parasite on the large corporation."*

Kelly envisioned the solid-state triode as a replacement of the vacuum tube in future telephone systems. He felt that mechanical relays in telephone exchanges systems caused problems and were very expensive to maintain, whence they might be replaced by some electronic device.

In the mid-1950s, American Telephone & Telegraph was the dominant provider of telecommunications services in the U.S. Through its system of companies, it owned or controlled 98% of all long distance telephone services and 86% of all facilities providing short distance telephone services. These operating companies bought most, if not all, of their equipment from Western Electric, the manufacturing branch of American Telephone & Telegraph. Western Electric produced telecommunications equipment based on the research carried out by the Bell Laboratories. American Telephone & Telegraph together with all the companies in the Bell System, employed 746,000 people with a total revenue of $5.3 billion or 1.9% of the U.S. GDP at the time. Between 1940 and 1970, Bell Laboratories filed on average 500 patents or 1% of all U.S. patents each year.

Bell Laboratories had better research resources than universities. Department managers, called department heads, played a role similar to the chairperson of departments at universities. Murray Hill had an excellent library, containing most of the latest technical books and journals. If a book was not locally available, the librarian would get it from another library. Researchers could buy all the books they wanted for their research at Bell Laboratories' expense.

Not immediately visible was the social structure of the organization. Bell Laboratories maintained a very casted social structure and people were very class conscious. There was a big difference between members of the technical staff and the support

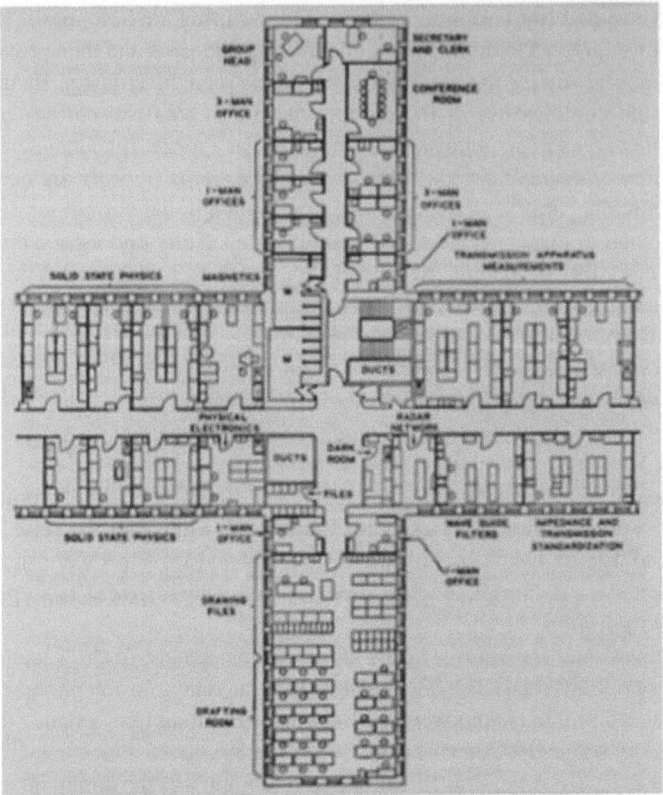

Fig. 6.4 The layout of the Solid-State Physics Department

staff. There were different dining rooms and it was just very conspicuous who you were eating with.

Considering what Bell Laboratories achieved, the reader may wonder why the phenomenon of the Bell Telephone Laboratories was never repeated. This book describes the invention of the transistor by Shockley's group, but Bell Laboratories produced a large number of other innovations, too. The silicon solar cell, a laser, the first communications satellites, the theory and development of digital communications, and the first cellular telephone systems. Bell Laboratories also built the first fiber optic cable and developed the Unix operating system and C computer language.

So how can we explain how a group of scientists and engineers, working at Bell Laboratories over a relatively short period of time, came out with such an astonishing range of new technologies and ideas? At Bell Laboratories, the man most responsible for the culture of creativity was Dr. Mervin Kelly. Born in rural Missouri to a working-class family and then educated as a physicist at the University of Chicago. His fundamental belief was that by "instituting a creative technology" and fostering a busy exchange of ideas between talented people, new discoveries would inevitably result. Kelly purposefully mixed together physicists, chemists, metallurgists, and electrical engineers; side by side were specialists in theory and experimentation who worked with Bell Laboratories' manufacturing plants to transform all these new ideas into reality.

Dr. Kelly believed that freedom was a crucial factor in research organization and he trusted people to create. Finally, there was another element that accompanied Mervin Kelly's innovation strategy, and that was his personal example. An element even more crucial than all the others. Throughout his life he kept to a highly regimented day that began with tulip gardening at 5 AM, having his driver take him to work at high speed, and after work reading until midnight. Dr. Kelly talked fast and walked fast; he ran up and down staircases. He despised the "yes" men and he let everyone know where he stood.

Very likely there is no better way to innovate.

Chapter 7
The Nuclear Reactor

> *"I do not believe in excuses. I believe in hard work as the prime solvent of life's problems."*
>
> W.B. Shockley

Prior to the 1930s, it was considered impossible to "split" an atom. By 1939, Otto Hahn and Fritz Strassmann had discovered that neutron bombardment of uranium produced an isotope of barium. Lise Meitner and her nephew Dr. Otto Frisch actually confirmed Hahn's discovery and correctly interpreted the results as evidence for nuclear fission. Frisch confirmed this experimentally on January 13, 1939, proving that uranium was indeed split by neutrons, as well as providing mathematical proof. The greatest aspect of Meitner's work was the justification of the production of a huge quantity of energy duringfission.[1] In 1939, much of the physics community was involved in the growing advances toward fission made by European scientists. In the United Kingdom, German refugees Otto Frisch and Rudolf Peierls made a breakthrough, under professor Marcus Oliphant, indicating that it would be possible to make a bomb from purified U-235.

Niels Bohr arrived in New York on January 16, 1939 to attend a Conference on Theoretical Physics at the George Washington University. Frisch and Meitner had not wanted to tell Bohr about their discovery until he got on the boat because they knew that he would be unable to keep the secret. Bohr, eager to share the news with Enrico Fermi, immediately made his way to Columbia University. Brattain, Shockley, Fisk, and other members of the Bell Laboratories group attended Bohr's seminar. Later that year Franklin Roosevelt called on Lyman J. Briggs who was a director of the National Bureau of Standards to head the Uranium Committee to investigate the fission of uranium, as a result of the Einstein–Szilárd letter. In the letter they warned President Roosevelt that Germany might develop an atomic bomb: *"Germany has actually stopped the sale of uranium from the Czechoslovakian mines which she has taken over. That she should have taken such early action might perhaps be understood*

---

[1]L. Meitner, O.R. Frisch, "Disintegration of Uranium by Neutrons: A New Type of Nuclear Reaction". Nature 143 (3615), 1939, p. 239.

© The Author(s), under exclusive license to Springer Nature Switzerland AG 2021
B. Lojek, *William Shockley: The Will to Think*, Springer Biographies,
https://doi.org/10.1007/978-3-030-65958-5_7

on the ground that the son of the German von Weizsäcker is attached [to] the Keiser-Wilhelm-Institute in Berlin where some of the American work on uranium is now being repeated."

Briggs asked Mervin J. Kelly, Director of Research at Bell Laboratories, if he could get someone from Bell Laboratories to see whether or not fission could be used as a power source. Partly out of scientific curiosity, Kelly asked W. Shockley and his friend J. Fisk to do this. After graduation, Fisk had gone first to the University of North Carolina at Chapel Hill, N.C., then grew unhappy with that and wanted to leave. Shockley got Fisk into Bell Laboratories and he was first given a post under physicist J. B. Johnson.[2] Shockley and Fisk soon discovered that the answer to Kelly's question was affirmative, and in the process of researching the subject they actually figured out how to make a nuclear reactor.

Shockley came up with the idea: *"If you put the uranium in chunks, separated lumps or something, the neutrons might be able to slow down and not get captured and then be able to hit the U-235."*

James Fisk said in an AIP interview [3]: *"We worked very hard on it. And we applied for patents on it just in the normal course of events, and we were thrown out of court on every conceivable count after the war. And the reason, I think, was that Fermi and Szilard had had essentially the same idea and probably at about the same time. We may have been earlier, they may have been earlier, I don't know. I don't think that anybody will ever know. But they were working hard on this and we were doing this simply as an exercise to answer the question.*

And we went up to see Fermi one day, and incidentally there's a little discussion of that in a paper that Bill Shockley is publishing shortly. We went up to see him one day, and I think he was a little suspicious of us. The secrecy surrounding this was fantastic. And I think that Enrico felt that there wasn't any real point in making this kind of a calculation now. What you have to do is know what the cross-sections are. But we thought we knew them well enough to know whether you were in the right ball part."

Shockley and Fisk knew that the key point was to calculate the high capture cross-section area for slow neutrons in uranium 238. They had to make some approximations because there was no way to handle it otherwise. What was surprising was that Shockley's simplified calculations of nuclear cross-sections gave quite accurate dimensions for pile structures. In several months, Shockley and Fisk designed the world's first nuclear reactors (Fig. 7.3).

J.B. Fisk and W. B. Shockley reported to M. J. Kelly concerning uranium as a source of power (July 16, 1940) (Fig. 7.2).

Abstract:

The question "can nuclear energy be made available by use of the fission process" is investigated. An analysis of the problem is made in which a mixture of uranium and water or paraffin is considered. The necessary conditions for a chain reaction to take

---

[2]Johnson presented his theory of thermal noise at the American Physical Society meeting in 1928.

[3]Interview of James Fisk by L. Hoddeson, AIP June 24, 1976.

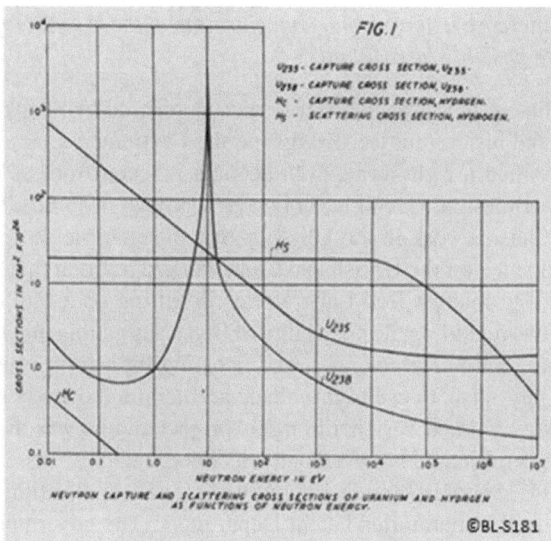

Fig. 7.1 Shockley and Fisk scattering cross-section calculations

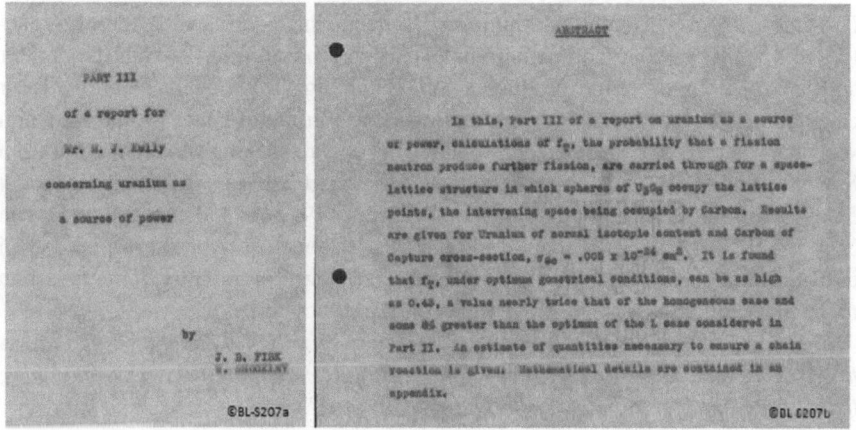

Fig. 7.2 Part III of Shockley and Fisk's report to M. Kelly (October 1940)

a place are obtained in terms of U, the number of atoms or uranium per atom and C_5 the isotopic U_{235} in the Uranium, for several values of v (v is the number of neutrons released per fissure.) Optimum ranges in which to work are determined together with minimum amounts or uranium and minimum total bulk of material necessary. The chief conclusion to be drawn from these calculations is that a sufficient quantity of uranium mixed homogenously with water, a continuing reaction may be possible *with only slight increase in the normal concentration of the U_{235} isotope; i.e. nuclear energy may quite possibly be available.*

Kelly immediately forwarded all three reports prepared by Shockley and Fisk to Lyman Briggs, and his committee. Briggs required extreme secrecy measures. The government classified it right away, even keeping it secret from its own scientists. Kelly later got permission to send it via Fowler who had been Fisk's professor and was then doing liaison work in the US with British scientists. It apparently had a considerable influence on the British and Canadian programs at that time.

Shockley tried to interest Bell Laboratories in setting up a trial unit. He sent a letter to Merle Tuve regarding the possibility of Bell Labs getting into nuclear physics by acquiring some kind of accelerator, and debating the various merits. Tuve was among the first physicists to use high-voltage accelerators to study the structure of the atom. Shockley worked very hard on this project and he was disappointed that his suggestion was evaluated but eventually declined.

In January 1941, Shockley and Fisk submitted a patent application titled "Nuclear Reactor" to the Bell Laboratories Patent Department. The government authorities, however, fought any attempt by Fisk, Shockley, or the Bell Laboratories to submit a patent (Figs. 7.3, 7.4 and 7.5).

A little-known fact is that Fermi in Chicago and Shockley and Fisk in Murray Hill independently invented the "lumping" principle that permits nuclear power reactors to operate without isotope enrichment when a mixture of the right moderator and natural uranium could produce a self-sustaining fission chain reaction. Their calculations proved that a large amount of natural uranium could produce enough secondary neutrons to keep a reaction going. There are striking similarities between Shockley and Fisk's patent application and Fermi and Szilard's patent application # 568,904 submitted to the patent office on December 19, 1944 and its patent issued on May 17, 1955. Perhaps the main difference was the enclosure of the reactor, Fermi used graphite blocks, Shockley and Fisk suggested paraffin (mixture of hydrocarbon molecules containing between twenty and forty carbon atoms).

Fisk later stated [3] *"Actually, the work was very largely Bill Shockley's genius. As far as I'm concerned, I was riding along, whipping up the horses. The only thing that bothers me is that he never had any credit for it."*

After the war Bell Laboratories re-submitted the Shockley and Fisk application to the patent office. The Manhattan Project physicists learned of Shockley and Fisk's work only after WWII. All this was an embarrassment, because the government had not only classified it right away, but even kept it secret from its own scientists. Bell Laboratories had no connection with the Manhattan Project in any formal way at that time. After wrestling with the government bureaucracy, Oliver Buckley then president of Bell Laboratories, gave up, and assigned any rights that might exist under Shockley and Fisk's invention to the government.

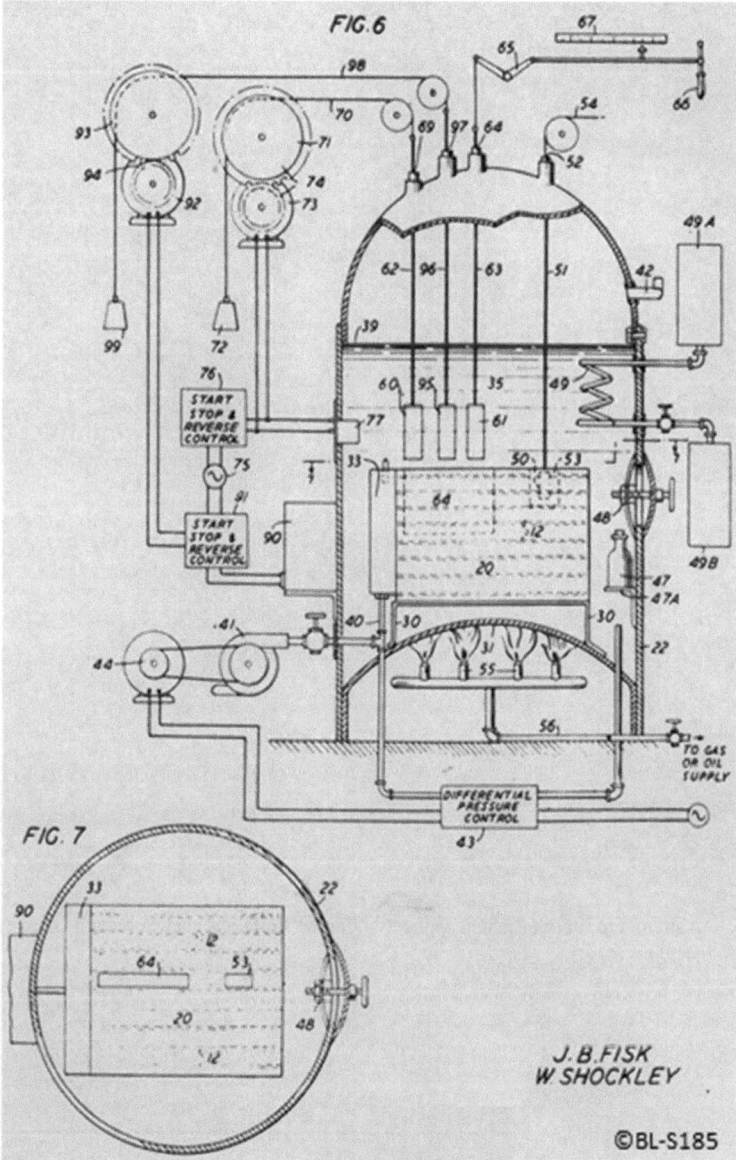

Fig. 7.3 Nuclear reactor designed by W. Shockley and J.B. Fisk in 1940

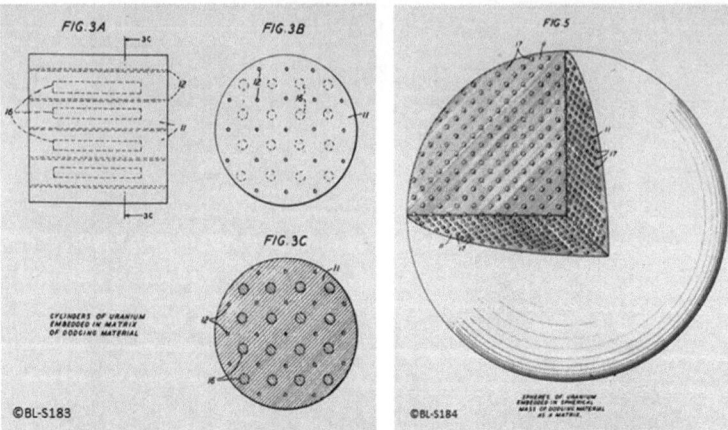

Fig. 7.4   Details of Shockley and Fisk's nuclear reactor

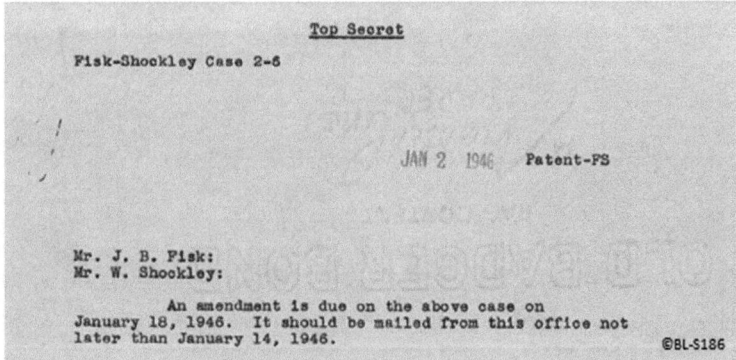

Fig. 7.5   An attempt to submit a patent application for the Bell Labs nuclear reactor (January 1946)

Chapter 8
World War II Hero

"Complete freedom is not very helpful to a person who is inexperienced in the world."

John R. Pierce (1910–2002).

In June 1940 President Roosevelt approved the request of Vannevar Bush to create the National Defense Research Committee (NDRC) *"to coordinate, supervise, and conduct scientific research on the problems underlying the development, production, and use of mechanisms and devices of warfare."* Members of the NDRC included: Dr. Vannevar Bush (chairman), president of the Carnegie Institution of Washington and formerly Dean of the faculty of engineering at MIT, Richard C. Tolman, Caltech, Irvin Stewart (NDRC Secretary), Rear Admiral Harold G. Bowen, Naval Research Laboratory, Conway P. Coe, U.S. Commissioner of Patents, Dr. Karl T. Compton, MIT, James B. Conant, Harvard University, Frank B. Jewett, president of Bell Telephone Laboratories, and Brigadier General G. V. Strong. The NDRC had neither the authority nor the funds to carry research forward into development and production. President Roosevelt issued Executive Order 8807 on June 28, 1941 establishing the Office of Scientific Research and Development (OSRD) as an independent entity with Vannevar Bush as director of the OSRD and given the authority to enter into contracts. J. B. Conant replaced Bush as director of the NDRC. Vannevar Bush designed his organization not only to mobilize the nation's civilian research and development capacity, but also to shield that capacity from government influence and bureaucracy.

Shortly after Christmas 1940, Dr. Jewett posted war posters in a hallway of Bell Laboratories and asked scientists to join the war effort (Fig. 8.1). Shockley moved to undertake radar research at the labs in Whippany, New Jersey. Shockley, 31 years old, knew his patriotic duty and he did not wait to be drafted. Captain Wilder Baker, head of the Antisubmarine Warfare Unit (ASW) at the First Naval District Headquarters in Boston was working closely with a British friend, F.M.S. Blackett, professor of physics at the University of Manchester. Blackett shared his opinion that the early-warning radar would not become fully effective until some of the civilian scientists and engineers went to the radar installations to work with military operators and convinced Baker to initiate a similar effort in the U.S. Baker contacted the NDRC

© The Author(s), under exclusive license to Springer Nature Switzerland AG 2021
B. Lojek, *William Shockley: The Will to Think*, Springer Biographies,
https://doi.org/10.1007/978-3-030-65958-5_8

Fig. 8.1 War poster at Bell Laboratories (1942)

with a request to find some civilian scientific help and suggested P. Morse, Shockley's favorite professor at MIT, as leader of the group.

Phillip M. Morse joined ASW in March 1942 and started recruiting. Morse's first cast was among his close friends, and W. Shockley was recruited first. By the end of 1940 the group had thirty members.[1] Shockley was director of research. The Commander of the U.S. Fleet, Admiral King insisted that Morse's group, which was paid by the NDRC, should work solely for the Navy and should not disclose any information to the NDRC without authorization by the Navy.

In November 1942 Shockley and Morse flew to London in a Pan American flying boat with refueling stops at Bermuda, Azores, and Lisbon. They arrived before the V-1 missile attacks began and immediately started working with Blackett. They learned how U-boat's operated and gathered information about the patrol flights by squadrons searching for surfaced U-boats. Morse returned to Washington after Christmas and Shockley stayed in London for another month. When Shockley returned to the States he analyzed all the available data. He found that on average the crew of an ASW plane had just one chance to find a German submarine during the crew's life before its members were killed. If an ASW crew did not engage correctly with a U-boat the first time, they usually had no other opportunity. Shockley and Morse started to apply the methods of operations research. The specific purpose

---

[1]One of the group members was Walter Widlar, the father of the legendary analog designer Robert Widlar.

of operations research in the war was to discover how to make the best use of the available military forces and weapons.

Shockley introduced the exchange rate *flying hours/# U-boats seen* and he found that this ratio ranged from 100 up to 25000. This analysis indicated that the geographical distribution of patrolling flights had to be changed. A new problem arrived when bombs dropped on spotted U-boats exploded and had little or no effect on the submarine. Shockley analyzed the ratio *tons of bombs on target/U-boat sunk* and very quickly found the problem. The dropped bombs were set up to explode at a depth of 75 feet. Such a deep explosion did not cause any significant damage when the submarine was at the surface. Shockley calculated that airborne bombs should explode at a depth of 35 feet. After the change was applied, the number of sunk submarines increased by a factor of five.

A few months after Morse's group was formed, the Navy consolidated all anti-submarine operation under the Antisubmarine Warfare Operations Research Group (ASWORG) and transferred the group to Washington, reporting to Admiral F.S. Low, Admiral King's deputy. The group was housed in the Navy building and was funded through a contract with Columbia University. Admiral King insisted on paranoid secrecy rules and only nine personnel with badges containing the letter "M" could enter the building. Other members of ASWORG were in England, Cuba, Iceland, Morocco, and other places.

With the submarine threat as a major issue Shockley's research was of paramount importance. As a result of his work he met many high-ranking officials and soon became involved in a variety of projects addressing different elements of the war effort. This involved devising methods for countering the tactics of submarines with improved convoying techniques, optimizing depth charge patterns, and so on. This project required frequent flights, often in uncomfortable "war conditions". In addition, Shockley and Morse were faced with the clash of military and scientific culture. Admiral King's paranoia required unnecessary bureaucracy, frequently preventing Shockley from taking time off. Shockley had very limited contact with his family and almost no opportunity to visit his family in Madison. Bill did not want Jean to be worried about his safety but because the strict war censorship did not allow him to tell his family where he was, Jean frequently did not hear from Bill for a long time.

By the end of 1943, Allied naval and air power had scored a victory over German U-boats, and submarines were dropped as the highest priority of Admiral Dönitz, a German naval officer and creator of the U-boat fleet.

In January 1944, W. Shockley was promoted to "Expert Adviser to the Secretary of War" with decision authorization at the level of Army General. Shockley was a trusted member of the War Department and with his security clearance he got rid of Admiral King's bureaucracy. His new assignment was to analyze the costs and benefits of bombing and prepare general training procedures for B-59 bombers equipped with MIT's new radar. Using the techniques of Operations Research, he assessed whether the cost to the U.S. in training aircrews, building airplanes and bombs, and losing a percentage of those on every raid, was justified by the cost to the German economy of the raids. Shockley moved his office to Smoky Hill Air Base in

Fig. 8.2 W. Shockley at
Smoky Hill Air Base in
Salina, Kansas (1944)

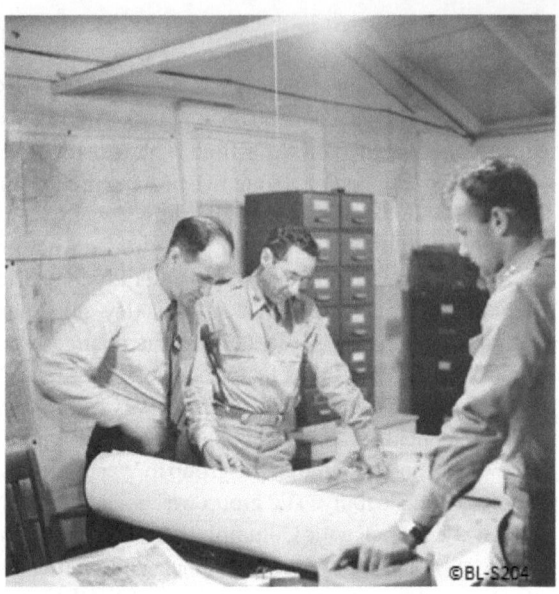

Salina, Kansas, where he began a series of flights taking pictures showing what the
operator might see on the radar screen alongside the actual image (Fig. 8.2).

In a very short period of time Shockley prepared a secret report and listed
several tactical mistakes he found in European war operations. He concluded that
the bombing of German cities had a smaller effect on the German war effort than
expected. Shockley wrote: *"The city bombing prior [to] autumn of 1944, did not
substantially affect the course of German war production. As a rule, the industrial
plants were located around the perimeter of German cities and characteristically
these were relatively undamaged."* A Survey Study completed after the war showed
that, overall, only about 20% of the bombs aimed at targets fell within a radius of
1000 feet around the target.

After working around the clock, on September 13, 1944, Shockley flew to England
to implement his training methods. These methods involving a high degree of
bombing precision were gradually adopted by many crews and an accuracy of 70%
was reached for the month of February 1945.

In November 1944 W. Shockley was dispatched to the Pacific military operations to evaluate the training and efficiency of B-29 bombers. He implemented
a new bombing tactic in which, instead of each bomber releasing its bomb load
independently of the others, all the aircraft released their bombs at the same time,
under the command of the lead bombardier (Fig. 8.3). At that time this strategy
was against regulations and forbidden, but no one in the Air Forces knew why. This
change resulted in a significant increase in the number of bombs on target. In August
1945, night bombing of the oil refinery in Shimotsu completely destroyed it and the
bombing accuracy of all B-29 crews was within 1700 feet of all targets.

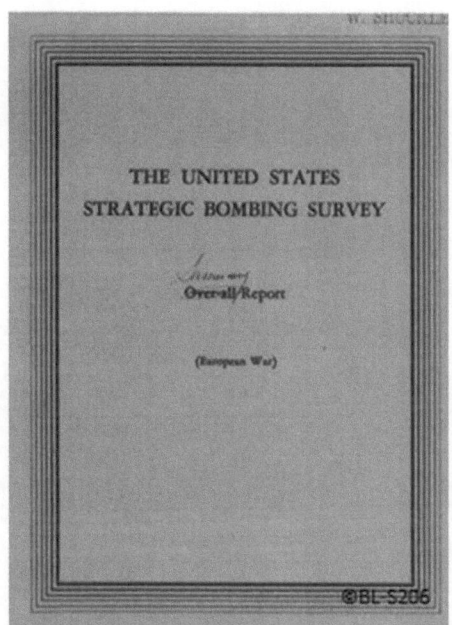

Fig. 8.3 Shockley's bombing strategy report

Fig. 8.4 4. H2X Scanning radar developed by Bell Laboratories, Western Electric, and MIT

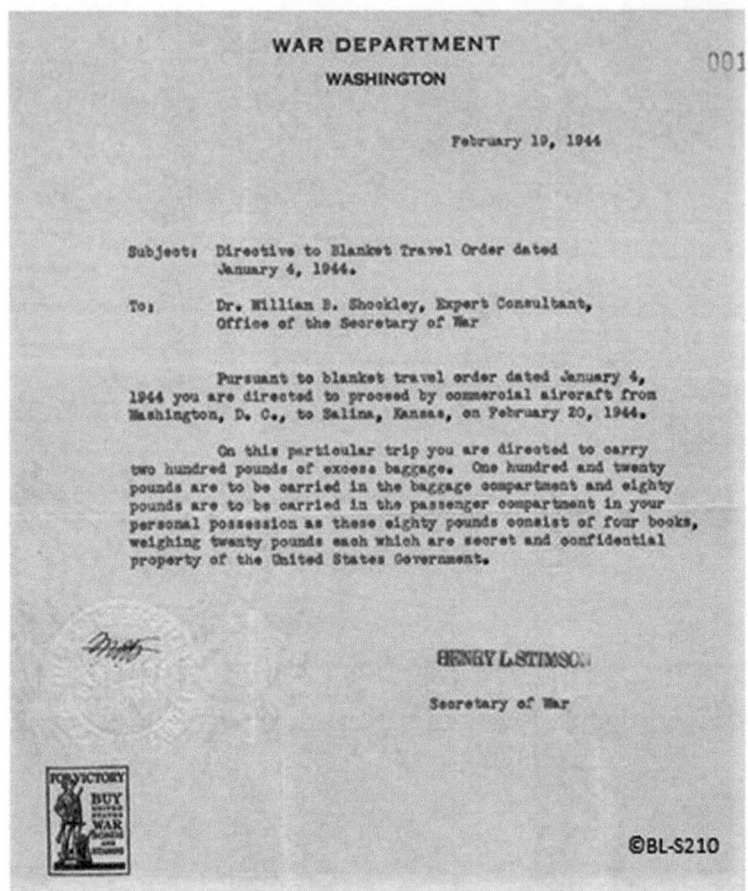

Fig. 8.5   Shockley's travel order dated February 19, 1944

Shockley was not treated differently than other commanders of the Pacific Theater. For his trip to the Pacific theatre he was given instructions as to the amount of personal belongings he could take with him (Fig. 8.5). He worked around the clock, moved from base to base in India, Australia, Ceylon, Saipan, living with sleep deprivation and under constant stress. During Christmas 1944, while in the trenches with other commanders, he was exposed for the first time to combat experience.

To introduce the B-29 into combat, bombers were based in India to strike Japanese targets in Indochina. Combat operations began on June 5, 1944, with the bombing of Bangkok, Siam (Thailand). In order to bomb Japan itself, Chinese staging bases were prepared. To mount a mission from China, the B-29 s had first to ferry their supplies from India over the "Hump" to China. When sufficient material was accumulated, the B-29 s struck Japan from their Chinese bases.

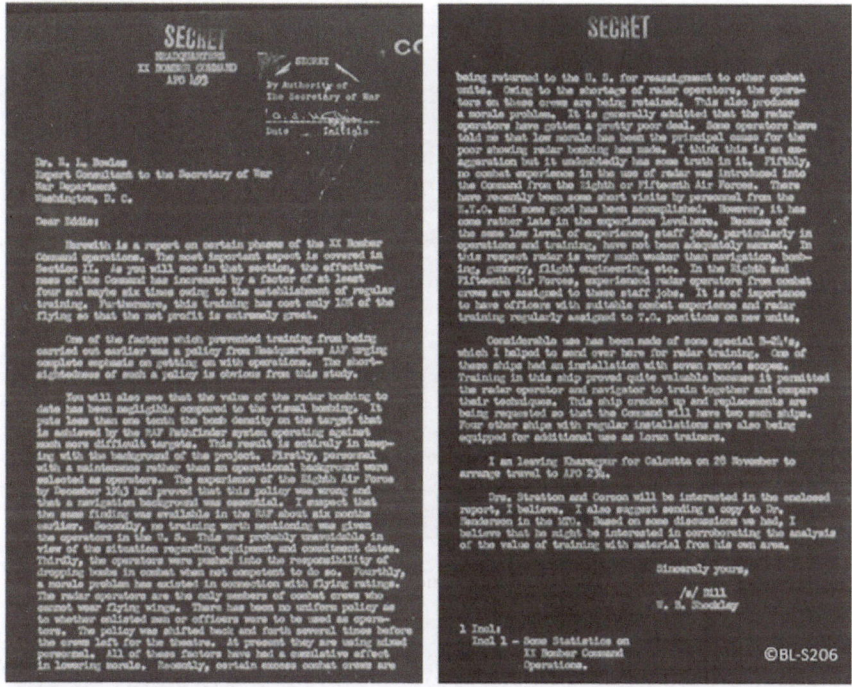

Fig. 8.6 Shockley's report to E. Bowles: *effectiveness of bombing has increased by a factor of at least four*

Shockley's methods of operational research applied to the cost-benefit analysis of aerial bombing were eye-openers to military management which had never before quantified the efficiency of aerial raids over Japanese targets Tokyo, Nagoya, Osaka, and Kobe (Fig. 8.6). The Air Force lost 5–10% of B-29 bombers in every mission, Shockley recommended that, rather than targeting Japanese cities, they should attack Japanese shipping convoys and military targets. In July 1945, the War Department asked Shockley to prepare a report on the question of probable casualties from an invasion of the Japanese mainland. In just one week, Shockley submitted a memo to Bowless [2] and stated *"the behavior of nations in all historical cases comparable to Japan's has in fact been invariably consistent with the behavior of the troops in battle."* The numbers from the last six months' operations indicated that, for every 10 Japanese soldiers dead, one U.S. soldier would die. In other words, the U.S. would probably have had to kill at least 5–10 million Japanese. This could have cost between 1.7 and 4 million casualties, with some 400,000–800,000 killed. This prediction influenced the decision for the atomic bombings of Hiroshima and Nagasaki to force Japan to surrender without an invasion.

---

[2]*"Proposal for increasing the scope of casualty studies,"* Office of the Secretary of War, August 1945.

Fig. 8.7   Secretary of War Robert P. Patterson and William B. Shockley

When the term "operational research" was first used in the late 1930s, nobody could have predicted the radical turns its history would take in World War II. These turns were guided by individual experiences and agendas of many institutions and people like Shockley. The methods of operational research used by the U.S. military today derive from Shockley's work during World War II.

As a result of Shockley's contributions, Secretary of War Robert P. Patterson awarded Shockley the Medal for Merit on October 17, 1946. At the ceremony with Allison in the audience William Shockley became one of the highest-ranking civilian scientists outside Los Alamos, and was the keeper of some of America's most closely held secrets (Fig. 8.7).

Speech to accompany the award of the Medal for Merit to Dr. W. B. Shockley:

Dr. William B. Shockley, for exceptionally meritorious conduct in the performance of outstanding services to United States since January 1, 1944. Dr. Shockley, as Expert Consultant in the Office of the Secretary of War, displayed great initiative, foresight, and ability of the highest order in advising the Army Air Force on training and operational techniques in the Very Heavy Bombardment Program in the China, Burma, India Theater of Operations, and in developing training facilities and methods for the improvement of radar bombardment. He served with distinction, exhibiting great tact and vision in initiating and perfecting operational policies and techniques which greatly improved the effectiveness of the Twentieth Air Force. Through unusual analytical facility and wide scientific contribution in the field of specialized operational research and analysis of broad military problems. By his tireless efforts, initiative and skillful application of scientific techniques to the problems confronting the Army Air Forces, he made an exceptional contribution to the war effort.

Harry S. Truman

The White House, July 19, 1946

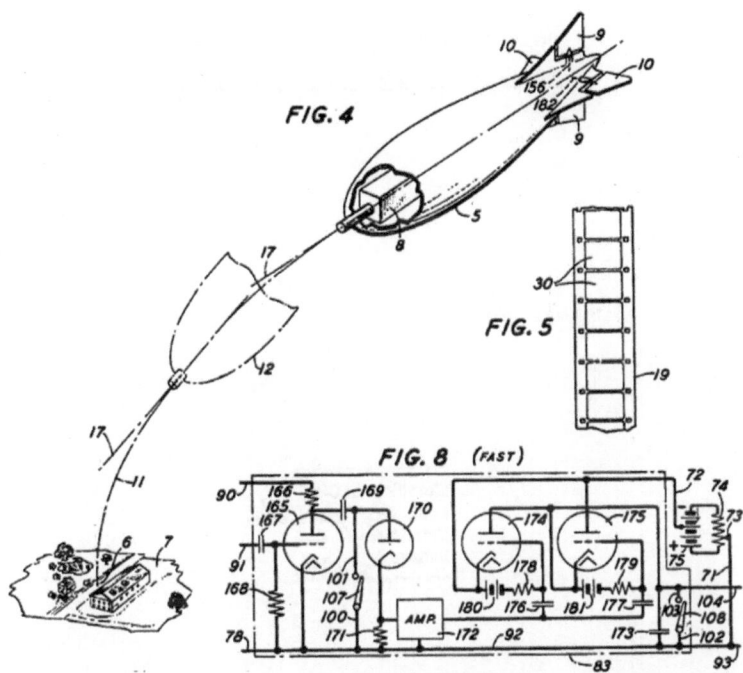

Fig. 8.8 A missile guidance system invented by W. Shockley (now U.S. Patent # 2,884,540)

After Shockley returned to Bell Laboratories the military requested continuation of his operations research work. When his security clearance expired, he was reappointed.

Wilber M. Brucker, Secretary of the Army:

> In recognition of his ability and patriotism Dr. William Shockley is hereby reappointed a member of the Secretary of the Army's Scientific Advisory panel, effective this date (August 12, 1956). The Department of the army seeks his counsel in the field of science and matters related thereto.

Shockley's involvement in work for the Navy and Air Force resulted in several inventions. On April 17, 1942 W. Shockley and G.W. Willard submitted a patent application "*Wave propagation device*" describing a system for the location of naval vessel and submarines. Shockley's prior work on electron multipliers resulted in the invention of a missile guidance system to "provide an improved system of control dependent upon the matching of patterns having two-dimensional characteristics" (Fig. 8.8). The invention was classified during the war, and Bell Laboratories submitted a patent application on March 19, 1948 after FIAT Reports[3] discovered that the Germans had designed a similar system.

---

[3]Field Information Agency Technical Reports (FIAT) were the U.S. Army's way of documenting, after WW II, the German advancement of science and technology suitable for proper exploitation by U.S. armed forces.

RELATIVE TECHNOLOGICAL ACHIEVEMENTS
IN
WEAPON CHARACTERISTICS IN USSR AND USA

DR. W. B. SHOCKLEY

30 JANUARY 1946

Classified SECRET
Auth: *off. Just. Her*
Date: *30 Jan 46*
Initials: *a.m.g*

SECRET   ©BL-S176

Fig. 8.9 Front page of Shockley's *Sputnik I* warning (January 1946)

During his involvement with the Air Force, Shockley invented an altimeter for accurately measuring aircraft altitudes. The instrument was basically a Wheatstone bridge exploiting changes in air pressure caused by the air flow through a narrow tube opening into a large vessel (U.S. Patent 2,509,889).

In January 1946, Shockley prepared for Pentagon Project 19 a report titled "Relative Technological Achievements in Weapon Characteristics in USSR and USA" (Fig. 8.9). By analyzing all available data, he argued that the Russian lag was somewhere between 0.5 and 1.1 years in 1944, and gave the following warning: "*During the war the Russia was under attack and [a] certain portion of her research and development establishment was destroyed. This has undoubtedly had an effect on increasing her lag as of 1944 and 1945. For this reason, the figures quoted probably overestimated her lag.*"

In June 1950, Edward Bowles, with whom Shockley had worked on the B-29 radar project during WWII, called Shockley and once again asked for help. Aging B-29 bombers that were not part of the nuclear strike force were released for combat in Korea. Many of these B-29 s were war-weary and brought out of five years of storage. Because of this, the U.S. suffered large numbers of casualties and lost aircraft (the total lost was 2,714 aircraft). In September 1950 Shockley flew to Korea with a group of military advisers to evaluate war operations. After a thorough analysis of the ground war, Shockley found old bombers were only part of the problem. The mortar shells being used were ineffective. After his recommendations were implemented, the Army used shells with proximity fuses, similar to those used in the Pacific theater.

As deputy director of the Weapons Systems Evaluation Group, Shockley was advising the Joint Chiefs of Staff. In January 1955, he prepared an ambitious proposal for a continental air-defense system to upgrade Nike-Hercules missiles and evaluate new accurate radar systems. Shockley resigned from WSEG on July 15, 1955 but continued to serve as Scientific Advisor for the Policy Council of the Joint Research and Development Board.

Following up on his 1946 Project 19 Report, Shockley describes how in 1957 the National Research Council and the Air Research and Development Command were conducting a study which had to do with long-range planning for the Air Force. They were so fearful of negative public reaction to the use of the word "space" that the word "exosphere" was used. The space projects had come up for such fierce congressional criticism that the military backed away from words that might reveal that they were interested in space. At this time, the Russians published an article in one of their journals clearly describing their satellite plan. Shockley states: "*After Sputnik I, there was a great deal of discussion about how we were falling behind Russia in science. The real problem did not have so much to do with science; it had to do with administration policy. Unfortunately, it took Sputnik I to bring about a realistic view.*"

When V. Bush, W. Shockley, and other scientists started working for the armed forces, the government was not a significant funder of science. Scientists were instrumental in the change which linked research, industry, and government to create a technocracy of expertise that governed Federal Science Policy after World War II.

Chapter 9
The Precursors of "Translating Apparatus"

"To understand science, it is necessary to know its history."

Isidore Marie Auguste Francois Xavier Comte (1798–1857).

Research at Bell Laboratories which resulted in the invention of the *translating device*, the term used before the name "transistor" was suggested, originated from an accumulation of observations and knowledge built up by several generations of scientists. The current state of science is always the result of an evolutionary process based on ideas of previous generations of scientists.

Faraday observed the decrease in resistance with increasing temperature of silver sulfide almost two hundred years ago. Ferdinand Braun demonstrated rectification, which was characteristic of a junction between certain metallic oxides or sulfides and a metal. Braun was studying the characteristics of electrolytes and crystals that conduct electricity. While probing a galena (lead sulfide) crystal using the point of a thin metal wire, Braun noticed that the current flowed freely only in one direction. In a paper[1] titled *"Über die Stromleitung durch Schwefelmetalle"*, submitted on March 1, 1883 to Annalen der Physik, Braun described the discovery of the rectifier effect. This property exhibited at the point of contact is referred to as the point-contact rectifier effect (Fig. 9.1).

This effect was exploited by many experimentalists studying electricity after 1900 who recognized that compounds, such as copper oxide, zinc oxide, and silver sulfide, do not conduct electric current in the same way as metals. By the end of the 1930s the theory of rectification had become quite evolved, although influential physicists, including W. Pauli and I. Rabi, considered semiconductors[2] like "dirt." L. Landau shared a similar opinion in a controversy with Abram Ioffe. The main problem was

---

[1] F. Braun, *"Einige Bemerkungen über die unipolare Leitung fester Körper"*, Annalen der Physik, Vol. 225 (1883), pp. 340–352.

[2] The word *"halbleiter* (semiconductor)" was introduced in 1910 by Josef Weiss in his doctoral thesis supervised by Johan Goerg Koenigsberger at the University of Freiburg, Breisgau. Koenigsberger opposed Weiss' terminology and never used the word *"halbleiter"* in his publications.

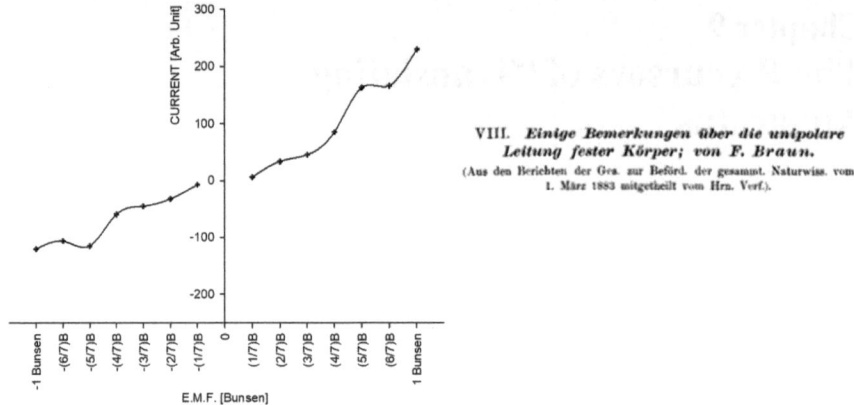

Fig. 9.1 F. Braun's demonstration of the rectifying effect (1 Bunsen is about 1.9 V). The arbitrary unit on the Y-axis is the number of divisions on a compass galvanometer

that the work concentrated on the defect semiconductors. Silicon or germanium were not considered to be semiconductors until 1930, when Bernhard Gudden published a review of the conductivity of semiconductors. He proposed that no chemically pure material would be a semiconductor. Gudden correctly considered that most semiconductor behavior was due to impurities (Fig. 9.2).

Copper monoxide Cu_2O, was the first semiconductor material investigated more thoroughly because of the industrial application of Cu_2O as a rectifier.[3] Properties of defect semiconductors such as copper oxide or zinc oxide depended to a large extent on the method of preparation. A rectifier made from the same copper or zinc could show a considerable spread of parameters, depending on the heat treatment or a mechanical or chemical treatment of the surface. For this reason, experimental data produced by different investigators seldom agreed with theoretical models.

Both Gudden and his teacher R. Pohl were obsessed with the accuracy and detail of their experiments. In 1924[4] while still at the University Göttingen, Gudden published his first detailed review of the electrical conductivity of crystalline substances, excluding ordinary metals. In 1930, he published a new article on the electrical conductivity of semiconductors.[5] He gave rather complete descriptions of results obtained for ionic or "electrolytic" conductors.

However, as far as semiconductors were concerned, he was unable to make real progress in understanding their properties because of the ambiguity in the experimental data. To illustrate this situation, Fig. 9.2 shows Gudden's comparison of

---

[3]L. O. Grondahl, P. H. Geiger, *"A New Electronic Rectifier,"* A.I.E.E. 46, 357–366 (1927).

[4]B. Gudden, *"Elektrizitätsleitung in Kristallisierten Stoffen unter Ausschluss der Metalle,"* Ergeb. Exakten Naturwiss. Vol 3 (1924), pp. 116–159.

[5]B. Gudden, *"Über die Elektrizitätsleitung in Halbleitern,"* Sitzungberichte der Physikalisch-medizinischen Sozietät zu Erlangen, Vol. 62 (1930), pp. 289–302.

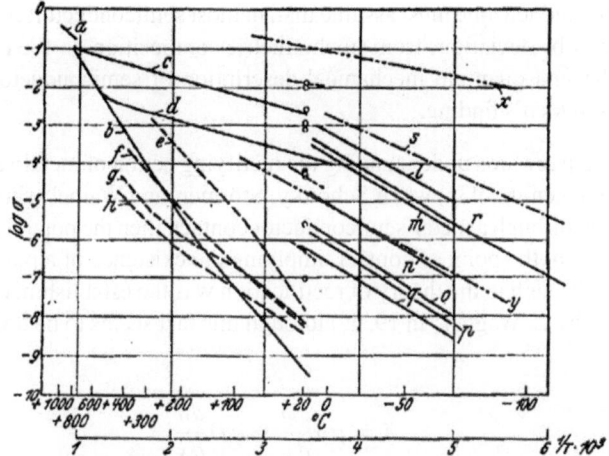

Abb. 2. Die Temperaturabhängigkeit der spezifischen Leitfähigkeit von Cu₂O.

a   WAGNER (für konstanten Sauerstoffgehalt berechnet); Kristallit.   Sauerstoffdruck 1 mm Hg.
     (Berichtigung: a ist um 4 mm nach oben zu verschieben.)
b   JUSÉ und KURTSCHATOW, ohne analytisch festzustellenden O-Überschuß; Kristallit.
c, d Dieselben, mit analytisch ermitteltem O-Überschuß von 0,1 und 0,06 Gewichts-%; Kristallit.
e   LEBLANC und SACHSE, Pulver der Zusammensetzung Cu₂O₁,₀₀₁₁.
f   Dieselben, Pulver der Zusammensetzung Cu₂O₁,₀₄₆.
g, h Dieselben, Kristallite.
l—q ENGELHARD (übereinstimmend mit VOGT), Kristallite verschiedener Vorbehandlung. (Auswahl aus
     rd. 100 Proben).
r   ENGELHARD; Messung zeigt schwachen Knick bei etwa — 70° C (Beispiel für mehrere Proben).
s   NASLEDOW und NEMENOW; Kristallit; Neigung bleibt bis — 183° C $\left(\frac{1}{T} \cdot 10^3 = 11\right)$ erhalten.
O   WAIBEL, nebst Andeutung der Temperaturabhängigkeit - - -; Kristallite verschiedener Vorbehandlung.
x   KAPP und TREU; dünne CuO- (Oxyd!) Schicht auf Glas, höchster erreichter Leitfähigkeitswert.
y   Dieselben, Cu₂O aus CuO unter O₂ erschmolzen.

Fig. 9.2 The conductivity of Cu_2O as a function of temperature as reported by Gudden

the spread of experimental data for Cu_2O. Gudden compared numerical values for all measurements then available with his version of the temperature dependence of conductivity:

$$\sigma = AT^n \exp\left(-\frac{E_0}{k_B T}\right)$$

He reached the conclusion that no chemically pure substance would ever be a semiconductor. The observed properties were believed to be due entirely to impurities. The puzzling question was why most of the fundamental properties of the investigated materials depended on such small quantities of impurities.

Gudden and his co-workers G. Mönch, E. Engelhard, B. Schönwald, and H. Voll recognized that the presence of small quantities of impurities would supply the free electrons which affected the conductivity in various ways. They also pointed out that pure semiconductors behave as insulators and are rare as natural samples.

According to Gudden, one must assume that, in most semiconductors, conductivity is caused entirely by deviations from stochiometric composition. In 1931, Alan Wilson published the first quantum mechanical description of semiconductor solids, thus confirming Gudden's finding.

The first step towards understanding the rectifying action of metal-semiconductor contacts was taken in 1931, when Schottky, Störmer, and Waibel[6] showed that, if a current flowed through a metal-semiconductor contact, then the potential drop occurs almost entirely at the point of contact, implying the existence of a potential barrier.

The breakthrough in the theory of rectification was the establishment of the diffusion equation by C. Wagner[7] in 1931, although this fact seems to be little recognized today:

$$-j = \mu n \frac{d\phi}{dx} + qD \frac{dn}{dx}$$

The theory of semiconductor rectifiers was advanced by Davydov[8] in the Soviet Union, Mott[9] in England, and Schottky[10] in Germany. N. Mott applied the Wagner equation under the assumption that the concentration of donor is vanishingly small, which leads to spatially constant electric field in the surface-charge layer. Neglect of mobile charge in the surface-charge layer is not always well justified, especially in the forward-bias direction. At the same time W. Schottky further improved the theory, but no exact solution of the basic equations was given.

It has been proven that the Mott theory and its Schottky extension gives the wrong current-limiting mechanism and the wrong current-voltage relation in metal/semiconductor diode rectifiers. The main problem was that the semiconductor materials available during the 1930's were very dirty, and it was impossible to link a theory with experiments. Copper oxide is a defect semiconductor with a relatively high acceptor activation energy. Compounding the difficulty of any theoretical understanding was the problem of controlling the exact composition of these early semiconductor materials, which were binary combinations of different chemical

---

[6]W. Schottky, R. Störmer, F. Weibel, "*Uber die Gleichrichterwirkungenan der Grenze von Kupferoxydul gegen aufgebrachte Metallelektroden,*" Z. Hochfrequenztech. Vol. 37 (1931), p. 162.

[7]C. Wagner. "*Zur Theorie der Gleichrichterwirkung,*" Phys. Zeitschrift, Vol. 32 (1931) pp. 641–645.

[8]B. Davydov, "*The rectifying action of semi-conductors*", The Technical Physics of the USSR, Vol. 5 (1935), pp. 87–95.

B. Davydov, "*On the photo-electromotive force in semi-conductors*", The Technical Physics of the USSR, Vol. 6 (1938), pp. 79–85.

B. Davydov, "*On the Contact Resistance of Semi-conductors*", Journal of Physics, Moscow, Vol. 1 (1939), pp. 167–174.

[9]N.F. Mott, "*The theory of crystal rectifiers*", Proc. Roy. Soc. Vol. 171A (1939), pp. 27–36.

[10]W. Schottky, "*Halbleitertheorie der Sperrschicht*", Naturwissenschaften, Vol. 26 (1938), p. 843.

W. Schottky, "*Zur Halbleitertheorie der Sperrschicht- und Spitzengleichrichter*", Zeitschrift für Physik, Vol. 113, (1939) pp. 367–414.

elements (such as copper and oxygen) and which strongly depended on the method of material origin and preparation. Semiconductor theory at that time could not yet explain exactly what was happening to electrons inside these devices, especially at the interface between copper and its oxide.

The diffusion theory is based on the assumption that the free path of electrons is shorter than the thickness of the space-charge region. This is not necessarily true if the space-charge region is very thin. The correction was developed by Hans Bethe.[11] In Bethe's theory, the current is limited by thermionic emission of electrons over the metal–semiconductor potential barrier. However, Bethe's condition is satisfied only over a relatively limited range. Usually accepted values of the drift velocity and of the carrier density at the barrier maximum are not correct, the drift velocity being underestimated by a factor of 2, and the carrier density being overestimated by about the same factor. These discrepancies compensate in the product of the carrier density and drift velocity, and this yields the correct forward current.

On November 25, 1937 Davydov submitted an article titled "The rectifying action of semi-conductors" to the Journal of Technical Physics of the USSR, published in English. His paper was the most complete description of the rectifying contact and provided the details of the solution of the drift-diffusion equations with the correct form of the drift-diffusion equations, including generation and recombination.

The work of Boris Davydov on rectifying characteristics of semiconductors seems to have escaped notice until after the war, even though it was available in English-language publications. Davydov, working at the Ioffe Physico-Technical Institute in Leningrad, came up with a model of rectification in copper oxide that foreshadowed Shockley's work on p-n junctions more than a decade later. His idea involved the existence of a p-n junction in the oxide, with adjacent layers of excess and deficit semiconductor forming spontaneously due to an excess or deficit of copper relative to oxygen in the crystal lattice. Nonequilibrium concentrations of electrons and holes (positively charged vacancies in the valence band) could survive briefly in each other's presence before recombining. At that time the set of non-linear differential equations shown in Fig. 9.3. created problems with its solution. Davydov was able, with a clever linearization procedure and appropriately simplified boundary conditions, to find a solution for the general nature of the phenomenon. Using this model, Davydov successfully derived the current-voltage characteristics of copper-oxide rectifiers; his formula was essentially the same as the one that Shockley would derive a decade later for p-n junctions.

At the end of the 1930s, Bell Laboratories was investigating materials with a negative temperature-coefficient that could be used in the underground transcontinental cable as a temperature compensator for the positive temperature coefficient of metallic components. W. H. Brattain, G. L. Pearson, J. N. Shive, H. R. Moore, and C. B. Green working in J. A. Becker's group studied a mixed oxide of nickel and

---

[11]H. A. Bethe, *"Theory of boundary layer of crystal rectifier,"* MIT Radiation Laboratory Report 43/12 (November 23, 1942).

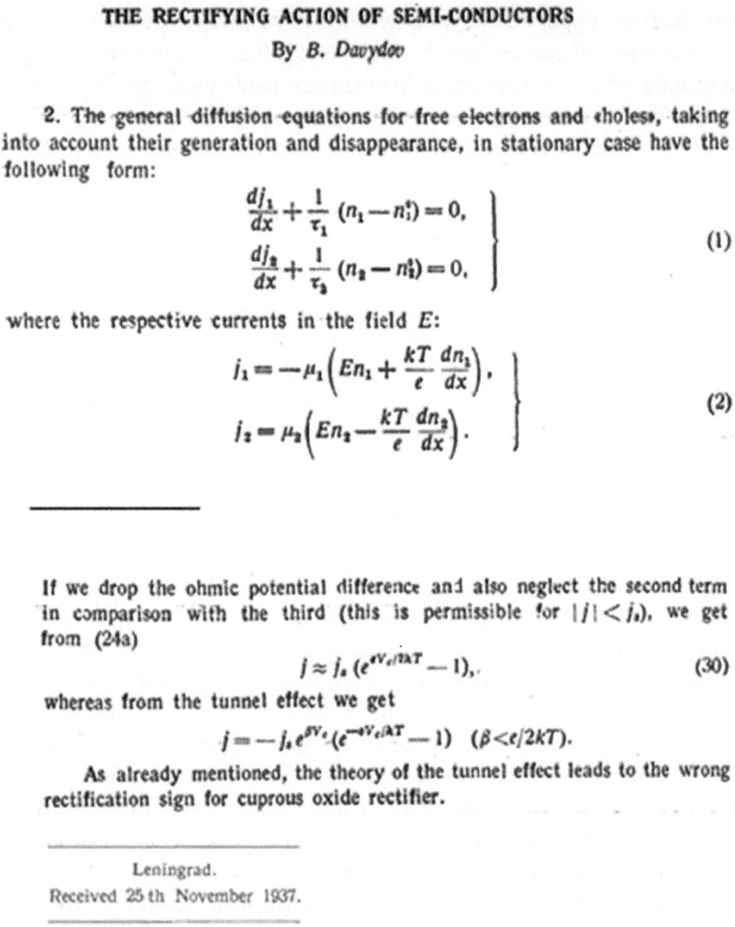

THE RECTIFYING ACTION OF SEMI-CONDUCTORS

By B. Davydov

2. The general diffusion equations for free electrons and «holes», taking into account their generation and disappearance, in stationary case have the following form:

$$\left. \begin{aligned} \frac{dj_1}{dx} + \frac{1}{\tau_1}\,(n_1 - n_1^0) &= 0, \\ \frac{dj_2}{dx} + \frac{1}{\tau_2}\,(n_2 - n_2^0) &= 0, \end{aligned} \right\} \tag{1}$$

where the respective currents in the field E:

$$\left. \begin{aligned} j_1 &= -\mu_1\left(En_1 + \frac{kT}{e}\,\frac{dn_1}{dx}\right), \\ j_2 &= \mu_2\left(En_2 - \frac{kT}{e}\,\frac{dn_2}{dx}\right). \end{aligned} \right\} \tag{2}$$

If we drop the ohmic potential difference and also neglect the second term in comparison with the third (this is permissible for $|j| < j_s$), we get from (24a)

$$j \approx j_s\,(e^{eV_e/kT} - 1), \tag{30}$$

whereas from the tunnel effect we get

$$j = -j_s\,e^{\beta V_e}(e^{-eV_e/kT} - 1) \quad (\beta < e/2kT).$$

As already mentioned, the theory of the tunnel effect leads to the wrong rectification sign for cuprous oxide rectifier.

Leningrad.
Received 25th November 1937.

Fig. 9.3 Snippets from Boris Davydov's paper (1937)

manganese compounds, which became known as the thermistor composition, manufactured by Western Electric. W. H. Brattain and his supervisor, Dr. J. A. Becker, conducted research on copper oxide and selenium rectifiers. But their work focused mainly on the manufacturing problems of copper oxide in Western Electric's plant.

Shockley was still in a vacuum tube department, running the Journal club, working on the nuclear reactor, and writing quantum mechanics lectures.[12] The first change occurred after Shockley studied the first published account of amplification using a solid state device, due to Hilsch and Pohl,[13] with a potassium bromide crystal (Fig. 9.4).

[12]W. Shockley "*The Quantum Physics of Solids*", BSTJ, Vol. 18 (1939), pp. 645–723.

[13]R. Hilsch, R. W. Pohl, "*Steurung von Elektronenströmen mit einem Dreielektroden Kristall mite in Modell einer Sperrschicht*", Zeitschrift für Physik, Vol. 111 (1938), pp. 399–408.

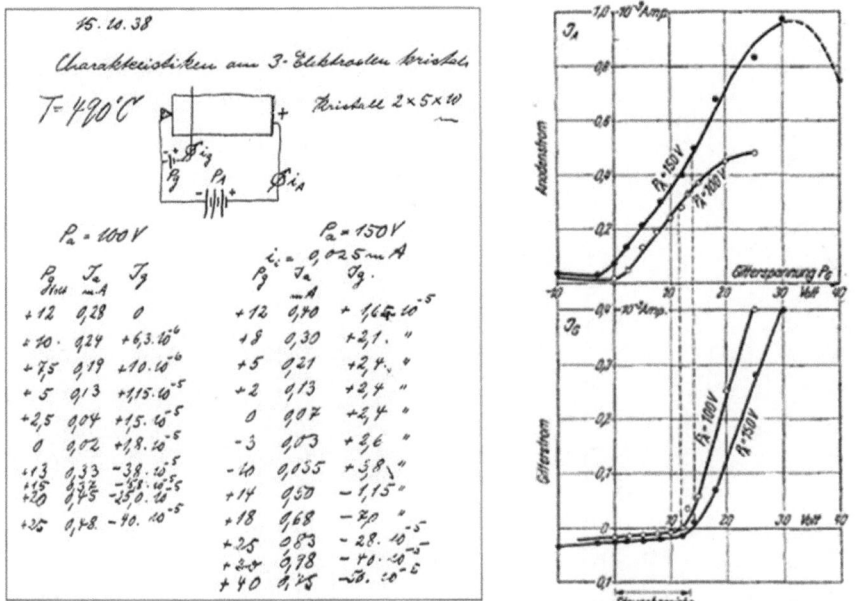

Fig. 9.4 A potassium bromide triode (Hilsch and Pohl, October 10, 1938)

Motivated by M. Kelly's thoughts about replacing the vacuum tube with a solid-state device and taking into account Hilsch and Pohl's idea, Shockley initiated his work on the copper oxide triode. He recognized the possibility of amplification by inserting a grid into the depletion layer which spreads more deeply into the semiconductor as the reverse potential on the rectifier is increased. Shockley saw that this spreading could be used as a kind of vacuum tube action to control the conductivity of the semiconductors. In the vacuum tube, electrons move unimpeded through the vacuum and can be influenced by the electrical field created by a grid. Shockley thought that the same mechanism might work in a cuprous oxide.

Shockley attempted to fabricate his triode by himself in the laboratory he shared with Nix. Dean Wooldridge described his effort like this: "*Shockley gingerly attached the fine mesh copper screen, apparently [...] cut out of some very old copper back porch screen with some dull scissors, to [the] green oxide coating and adjusted the voltage that applied to the mesh he hoped to control the current flow from one wire to the other. Of course, he was orders of magnitude away from anything that would work.*" But undaunted, Shockley approached Brattain with the idea that "*if we made a copper-oxide rectifier in just [the] right way, [...] maybe we could make an amplifier*" (Fig. 9.5).

Brattain said in an interview[14]: "*I had a good esprit de corps with him, and so after he explained, I laughed at him and told him that Becker and I had been all*

---

[14]W.H. Brattain in an interview with A. Holden and W.J. King, AIP June 1964.

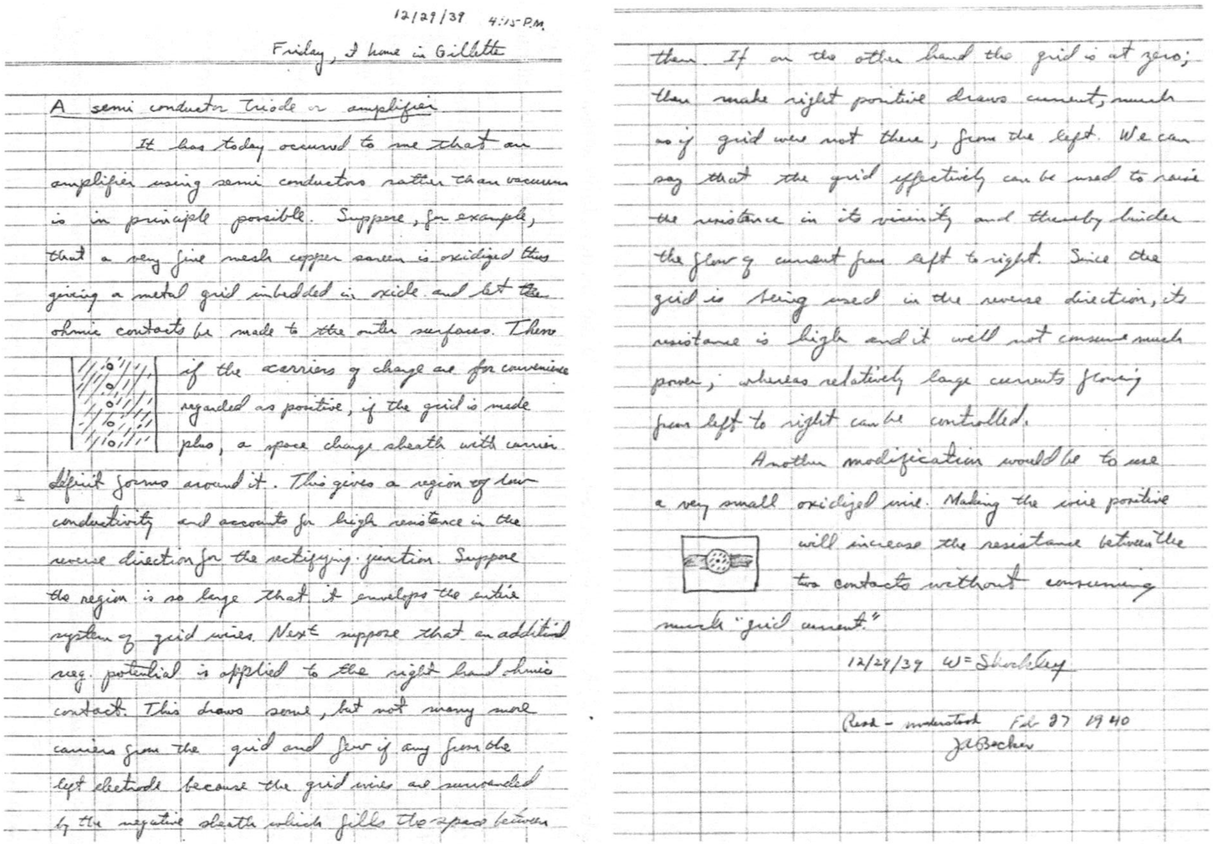

12/29/39 4:15 P.M.

Friday, I home in Gillette

A semi conductor triode or amplifier

It has today occurred to me that an amplifier using semi conductors rather than vacuum is in principle possible. Suppose, for example, that a very fine mesh copper screen is oxidized thus giving a metal grid imbedded in oxide and let the ohmic contacts be made to the outer surfaces. Then if the carriers of charge are for convenience regarded as positive, if the grid is made plus, a space charge sheath with carrier deficit forms around it. This gives a region of low conductivity and accounts for high resistance in the reverse direction for the rectifying junction. Suppose the region is so large that it envelops the entire system of grid wires. Next suppose that an additional neg. potential is applied to the right hand ohmic contact. This draws some, but not many more carriers from the grid and few if any from the left electrode because the grid wires are surrounded by the negative sheath which fills the space between them. If on the other hand the grid is at zero; then make right positive draws current, much as if grid were not there, from the left. We can say that the grid effectively can be used to raise the resistance in its vicinity and thereby hinder the flow of current from left to right. Since the grid is being used in the reverse direction, its resistance is high and it will not consume much power; whereas relatively large currents flowing from left to right can be controlled.

Another modification would be to use a very small oxidized wire. Making the wire positive will increase the resistance between the two contacts without consuming much "grid current."

12/29/39 W=Shockley

Read – understood Feb 27 1940
J.A.Becker

Fig. 9.5 Shockley's semiconductor triode (December 29, 1939 witnessed by J.A. Becker) *through this and that this sounded exactly like the same thing, and that I was quite sure it wouldn't work. But I said, "Bill, it's so damned important that if you'll tell me how you want it made, if it's possible, we'll make it that way. We'll try it."*

There are rather sharp discrepancies between Brattain's account and the written record. Brattain did not say what he and Becker proposed. None of the Becker and Brattain notebooks contain any information to explain what "the same thing" was and how it was tested. There are also no details regarding Shockley's triode and its testing by Brattain.

Two months later Shockley described the improved version of his "triode". The idea was that the current flowing between two electrodes on top of the copper oxide might be controlled by the potential of the metal electrode.

Many historians have created the impression that Lilienfeld invented the structure which was suggested by Shockley in February 1940 and which later resulted in the various versions of field-effect devices. Julius Lilienfeld filed an application for a patent in Canada and in the USA, describing a device which he then called "Method and Apparatus for Controlling Electric Currents."[15] Lilienfeld never built a working prototype of his device; indeed, there is no evidence that he ever even tried. He also never published any of his work other than patent applications. Lilienfield was well ahead of his time and this is the reason for the weakness of his device. The inferiority of the material system then available, and the technological inability to control materials and structures adequately is also very seldom made clear. What is not often recognized is that Lilienfield's and Shockley's device shown in Fig. 9.6 is what we call today a MESFET. Shockley's device envisioned later in 1945 is called a MOSFET in current terminology.

With America's entry into World War II, Bell Laboratories' work on semiconductors was interrupted and research concentrated on the war effort. In the early 1940s, wartime needs for point-contact rectifiers in microwave radar receivers became crucial, and extensive joint efforts with government and universities were undertaken. At Bell Laboratories, a group of metallurgists led by Schumacher and Scaff studied preparation methods and the properties of silicon. Although the first silicon ingots prepared by Scaff and Theuerer were polycrystalline and relatively impure, they revealed fundamental knowledge about semiconductors. This work on materials contributed greatly to the development of germanium point-contact rectifiers with better rectifying properties than silicon units, partly because of improved electrical contact forming techniques.

In one of the very early experiments, Russel Ohl and Jack H. Scaff melted commercially available silicon in helium ambient in a crucible of fused silica. The molten silicon was frozen in place and examined. Ohl found that the polarity of point-contact rectification reversed as the distance to the center of the ingot decreased. The study also showed that a rectifying barrier which, upon illumination, produced a photovoltage existed between the two regions. Scaff and Ohl named the outer region "p-type" and the inner region "n-type."

---

[15]US patent 1,745,175, submitted by Lilienfeld on October 8, 1926.

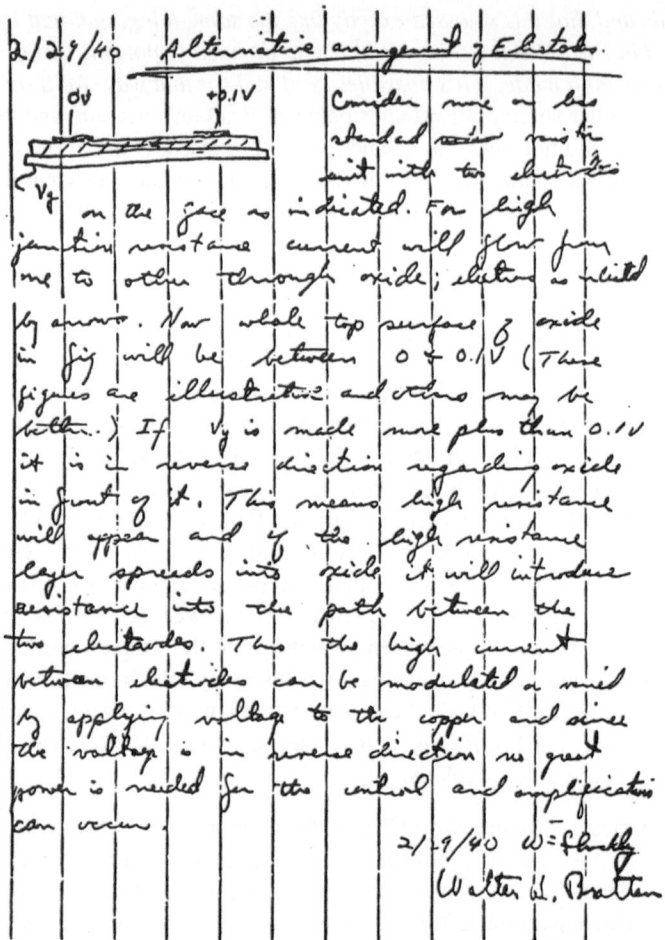

Fig. 9.6 Alternative arrangement of Shockley's Cu_2O triode (February 29, 1940)

The discovery of p-n junctions aroused intense interest at Bell Laboratories. W. G. Pfann and J. Scaff applied microscopy and special etching techniques to identify the p-n junction as a boundary separating the p and n regions of the ingot. These early ingots were badly cracked because they expanded when frozen and the silicon adhered to the crucible. Such ingots provided p-type silicon for the 1N21 series of point-contact rectifiers in 1942. The cartridge version of this unit was manufactured by Western Electric.

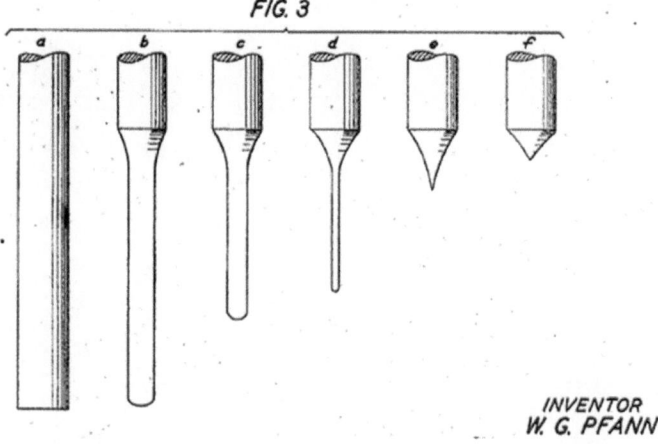

Fig. 9.7 Electrochemical thinning of point contacts invented in 1943 by W. Pfann

The results of their work were summarized in the paper of Scaff, Theuerer, and Schumacher[16] which was written before the end of 1945, but publication was delayed until 1949 owing to military security restrictions (Fig. 9.7).

A little known fact is that the key component, which enabled the invention of the point contact transistor, was Pfann's invention[17] of thinning of contacts. The critical part of the structure of translating or rectifying devices is the small contact point at the end of the contact wire which makes engagement with the surface of the rectifying crystal. A tungsten contact wire with a diameter of about 120 μm was originally ground and shaped to form a needle-like end. To prepare such a point contact was more art than science, but Pfann's invention of the electrochemical thinning of point contacts allowed production of reproducible point contacts.

---

[16]W. G. Pfann and J. H. Scaff, *"Microstructures of Silicon Ingots,"* J. Metals Vol.1, Trans. AIME 185 (June 1949), pp. 3089–3092.

J. H. Scaff, H. C. Theuerer, and E. E. Schumacher, *"p-Type and n-Type Silicon and the Formation of the Photovoltaic Barrier in Silicon Ingots,"* J. Metals, Vol. 1, Trans. AIME ll85 (June 1949), pp. 383–388.

[17]W.G. Pfann, Patent application, *"Method of forming a point at the end of a wire"*, Serial No. 498,376, dated August 12, 1943.

Chapter 10
The "Three Electrode Circuit Element Utilizing Semiconductor Material"

"Success has many fathers but failure is an orphan."

Count Gian Galeazzo Ciano, Diaries 1943

The discovery of the transistor at the widely heralded Bell Telephone Laboratories emerged as the subject of the study of many authors and historians. Detailed scrutiny of a large number of these works, especially works published after the advent of the internet, reveals inconsistencies, errors, and often the attribution of reality to what their authors wish to be true or a tenuous justification of what they want to believe.

The American Physical Society organized a "Symposium on the Solid State" in January 1945. Shockley discussed the conclusions of the symposium with the Director of Research at Bell Laboratories, Mervin J. Kelly, and his colleagues R. Bown and J. Fisk. Shockley said: "I *think semiconductor physics is the area to explore*" and argued that advances in solid-state physics and improved methods of producing semiconductor materials offered an opportunity for Bell Laboratories. At the end of the war, the research budget exceeded $60 million and Bell Laboratories employed about 2500 scientists, engineers, and technicians. Approximately 2000 technical employees were assigned to the radar project financed by the government.

On July 16, 1945 M. Kelly issued an "Authorization for Work" with the subject *"Solid State Physics—the Fundamental Investigation of Conductors, Semiconductors, Dielectrics, Insulators, Piezoelectronics and Magnetic Materials"*.

M. Kelly appointed James Fisk Assistant Director and organized three research groups: Physical Electronics managed by D. Wooldridge, Electron Dynamics managed by J. Fisk and Solid-State Physics with Shockley managing the Semiconductor group and chemist Stanley Morgan managing the Dielectrics, Insulators, Piezoelectronics and Magnetic Materials group.

William B. Shockley returned to Bell Laboratories in September 1945 after a brief vacation with his family in August. With Kelly's reorganization, a number of scientists moved into Shockley's group: W. Brattain (who in the meantime called his

© The Author(s), under exclusive license to Springer Nature Switzerland AG 2021   81
B. Lojek, *William Shockley: The Will to Think*, Springer Biographies,
https://doi.org/10.1007/978-3-030-65958-5_10

former boss J. A. Becker a SOB and did not want to work with him anymore)[1], experimental physicist G. L. Person, electrical engineer Hilbert Moore, chemist Robert Gibney, and technicians P. W. Foy and E. G. Dreher.

W. Shockley and S. Morgan's groups had no elemental semiconductor samples. They visited the Lark-Horwitz group at Purdue University on September 6 and 7, 1945. R. M. Whaley provided a review of Purdue's work on high resistivity germanium. R. Whaley had succeeded in producing high purity germanium in mid-1944 and provided the first few samples to Sylvania and the Bell Holmdel group which manufactured the 1N21 diode for MIT Rad Lab. Shockley sent a thank-you note back to Purdue stating *"our visit to Purdue was giving us concrete ideas to use in formulating our research program."* Shockley and Morgan decided that they would rely on the Purdue group for preparation of silicon and germanium. In a letter written to Lark-Horowitz just after the Shockley and Morgan visit, Whaley stated[2] *"They were much interested in all phases of our work. They have not yet formulated a program of research for their new Solid-State Division, but are accumulating information regarding problems and techniques in this field. Bell Labs did not know how to make high-back-voltage diodes and pure germanium alloys and about Purdue's theoretical work"*.

Shockley wanted to include a theoretical physicist in his group. A search inside Bell Laboratories did not yield anyone suitable. Shockley's friend, Frederick Seitz, recommended John Bardeen. Both Shockley and Fisk had met the humorless Bardeen when they were students, but they had no contact during the war. In the war years Bardeen had worked as a civilian physicist in the Naval Ordnance Laboratory and he told Shockley that he was returning to the University of Minnesota. Shockley asked M. Kelly to intervene, and Kelly used a trick which always worked: he offered Bardeen double his salary at the university, if he accepted Bell Laboratories' offer. Within one-month Bardeen and his wife moved to Summit, NJ and he joined Bell Laboratories on October 15, 1945.

Shockley revisited his 1939 concept of the copper oxide field-effect amplifier and suggested a modification of the device using a thin layer of germanium semiconductor. He hoped for a discovery of a purely electronic, rather than thermal, solid state amplifier.

On June 23, 1945 Shockley described a device, shown in Fig. 10.1, with a thin layer of semiconductor placed on an insulating support. Such a layer formed one plate of a parallel capacitor. The other plate was a sheet of metal placed in the proximity of the semiconductor layer.

Shockley calculated that a semiconductor layer thickness of 1000 Å with resistivity 2.4 Ωcm and mobility $\mu = 0.26 \, m^2/Vs$, should contain about 1×10^{21} electrons, giving 1×10^{14} electrons per square meter of surface. Then the electric field of 30 kV/cm and dielectric with dielectric constant 2 should produce

---

[1]Brattain complained *"Becker had a tendency that sometimes annoyed me to think, after he'd worked it out for himself, that it was his own idea."* [Brattain interview with A. Holden and W.J. King, AIP, June 1964].

[2]R. Whaley to K. Lark-Horowitz, letter dated September 10, 1945.

Fig. 10.1 1945 version of
Shockley's semiconductor
triode

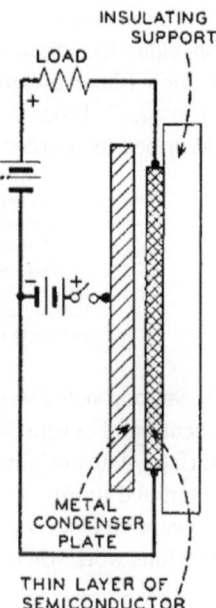

3.3×10^{14} electrons/m^2, or three times the number of electrons without an electrical field. Therefore, the conductivity of the semiconductor $\sigma = qn\mu$ should triple the conductivity of the semiconductor without the field.

Shockley's group still did not have a good crystalline semiconductor. G. L. Pearson set up the experimental verification with various p-type germanium (polycrystalline and evaporated), n-type silicon (polycrystalline), and copper oxide samples available and evaporated gold contacts on opposite sides of the semiconductor layer. Pearson found that the change in conductivity was only about 10%.

On November 7, 1945 Shockley asked Bardeen to check his calculations. Bardeen did not find anything wrong with Shockley's numbers. Neither he nor Shockley could explain why the experiment did not agree with the theory. On March 19, 1946 Bardeen wrote down in his notebook "*the electrons drawn to the surface of the semiconductor, when it was negatively charged, were not free to move as were the electrons in the interior. Instead, they were trapped in surface states, so they were immobile. Thus, in effect, the surface states trapped the induced charge of electrons and thereby shielded the interior of the semiconductor from the positively charged control plate.*" This was a great insight! Bardeen's empirical suggestion explained Pearson's experimental data. The modulation effects were far less than the prediction of theory, because a high portion of the induced charge trapped in the surface states prevented the addition of carriers to increase conductivity. By considering the Bardeen assumption it was possible to explain additional observations that did not agree with the Schottky theory, and why, contrary to expectation, the properties of a metal-semiconductor detector did not depend on the type of metal used. The problem was that Bardeen's insight did not suggest a method for mitigating this phenomenon.

The unanswered question is why Shockley himself did not suggest an explanation. Shockley had studied the surface states in 1939 and published a paper[3] in Physical Review where reference 8 is cited as: E.T. Goodwin, "*Electronic states at the surfaces of crystal*", Proc. Camb. Phil. Soc., Vol. 35 (1939), pp. 205–220, which clearly described the Bardeen finding:

> It is shown that in a crystal there exist states in which the electron is bound to a surface of the crystal and has an energy lying within a forbidden band. The wave functions and energies of these states are calculated, on the nearly free electron approximation, in terms of the constants of the crystalline potential field, which is represented by a triple Fourier series having the periodicity of the lattice. The method is shown to be applicable to a general crystal having a surface parallel to any one of the crystal planes.

Surface states were discovered by I. Tamm in 1932 and the topic attracted the attention of several authors. In his paper Shockley investigated how the energy of surface states originates from the discrete levels of atoms. Shockley's paper was submitted to Physical Review on June 19, 1939, more than six months before his conception of a copper oxide triode and obviously he did not think enough about his previous work when he suggested an improved version of the semiconductor triode in 1945.

Bardeen published his finding in 1947[4] with reference to both Shockley's and Goodwin's papers. He interpreted the surface states in terms of conductivity in the surface space-charge layer, due to band bending. What was not known then is that surface states alter the surface space-charge layer under the surface by attracting excess charge into the surface states; this must then be compensated by an equal and opposite space-charge layer. Such charging of the surface alters the energy band in the vicinity of the surface. In today's terminology, surface states are responsible for charge carrier accumulation, depletion, and inversion near the semiconductor surface.

The surfaces of the samples used in Pearson and Brattain's experiments were dirty and ill-defined. Their experiments were poorly defined, so an idealized picture of surface states could never apply to the surfaces of the samples they used. Brattain's notes do not provide any details about sample preparation or surface treatment of samples used in experiments, except that samples "*were ground flat with abrasive mesh.*" For the typical 400 grit finish they used, the RMS was about 0.25 μm. Such rough surfaces resulted in enormous surface charge and significant reduction of carrier mobility. Based on their data the surface state density was in order of $5 \times 10^{13} - 1 \times 10^{14}$ states/cm^{2}eV. This is approximately one surface state per ten surface atoms. Such material would not be considered today a semiconductor, but just dirt. The majority of the experiments were performed with evaporated layers (on quartz plates) or polycrystalline samples where the surface charge further increased.

---

[3] W. Shockley, "On the surface states associates with a periodic potential," Phys. Rev. Vol. 56 (1939), pp. 317–323.

[4] J. Bardeen, "Surface states and rectification at a metal semi-conductor contact", Phys. Rev. Vol. 71 (1947), pp. 717–726.

If Bardeen and Brattain had had samples with the well-defined surfaces states that are available today, they would have discovered the importance of surface conductivity rather than band bending. In the early Bardeen and Brattain experiments, the surface-state conductivity was not taken into account because they assumed (without real evidence) that surface-state conductivity would be negligible. They were very likely motivated by the fact that the band bending could be varied in a reproducible way by exposing the surface to different gaseous ambients.

Bardeen's March 1946 observation was a major breakthrough, Shockley wrote: *"This new finding was electrifying. Thus, the semiconductor research group had some new and exciting physics to work on. We abandoned the attempt to make an amplifying device and concentrated on the new experiments related to Bardeen's surface states".*

The group's enthusiasm and intense interest in the research worked well. They met daily and compared notes and results of experiments. They shared information freely, Shockley usually at the blackboard presenting new ideas, and then Brattain complaining: *"I will bet a dollar it would not work."* Brattain stated later [5] *"I gave up because he was winning too often."*

The possibility of the existence of surface states was the subject of considerable discussions in the literature, but there was almost no direct evidence for them. Shockley came up with an idea for an experiment which might provide such evidence. Because the Fermi level in semiconductors changes with doping level, he speculated that the contact potential of p-type and n-type should change. H. Moore, and E.G. Dreher designed modified instruments for contact potential measurement developed at the University of Pennsylvania by W. E. Stephens and W. E. Meyerhof [6]. The major modification was a biased platinum mesh reference electrode which allowed the sample to be exposed to a chopped illumination. The instrument was completed by the end of May 1947, and in June, Brattain demonstrated the evidence for the surface states from the change in the contact potential with and without illumination. Brattain and Shockley published their results in Letters to the Editor of Physical Review in July of 1947 [7,8]. The present author rates reference 7 as the worst Shockley publication. The usually crisply clear and precise Shockley publications contrast sharply with this submission, where almost nothing is clear.

The proof of the existence of surface states triggered a large number of experiments. According to Bardeen's suggestion, positively charged surface states would reveal the majority of carriers in p-type germanium. If the sample was exposed to light, generated hole-electron pairs would be separated by the electric field at the

---

[5]BTL Dinner party "25th anniversary of transistor invention", Murray Hill, NJ, December 25, 1972.

[6]W. E. Stephens, B. Serin, and W. E. Meyerhof, *"A Method for Measuring Effective Contact e.m.f. between a Metal and a Semi-conductor,"* Phys. Rev. Vol. 69 (1946), p. 42, and p. 244.

[7]W. H. Brattain, W. Shockley, *"Density of Surface States on Silicon Deduced from Contact Potential Measurements"*, Phys. Rev. 72 (1947), p. 345.

[8]W. H. Brattain, *"Evidence for Surface States on Semiconductors from Change in Contact Potential on Illumination"*, Phys. Rev. 72 (1947), p. 345.

sample surface: electrons would be attracted by the positively charged surface states and holes would move towards the grounded sample electrode. Such carrier separation would result in an increase in the negative surface potential. Bardeen was a theoretician and although he had studied electrical engineering, he was clumsy and could not do anything with his hands. Brattain and Pearson performed all experimental work. In the course of Brattain's experimental work, Bardeen suggested that trapping of electrons in surface states should be temperature dependent and asked Brattain to perform the experiment. Brattain did not have a thermostat, so he inserted the header with his samples into a thermos bottle. When he warmed the sample to higher temperature, the walls surrounding the sample were the hottest and the sample the coolest. The moisture in the air condensed on the semiconductor sample and Brattain noticed changes in the contact potential.

To avoid condensation, Brattain filled the thermos bottle with distilled water, then with ethyl alcohol. He observed similar changes in the contact potential regardless of the liquids he used. He had no explanation for such an effect and asked R. B. Gibney, a physical chemist, for help. Gibney realized that the condensed water or liquid in the thermos bottle created an electric field between the biased platinum electrode (Brattain maintained the constant bias of the reference electrode at $+14$ V) and the sample surface. Gibney asked Brattain to change the potential of the platinum reference electrode. They found that a positive potential on the reference electrode increased the contact potential, while a negative potential reduced it.

Gibney made a critical discovery that the voltage applied between the reference electrode and semiconductor sample, while both were immersed in electrolyte, created the electric field perpendicular to the sample surface and increased the electric field inside the semiconductor sample.

On November 17, 1947, Brattain recorded in his notebook the page shown in Fig. 10.2. Gibney's idea of using electrolytes to overcome the blocking effect of the surface states energized the group's search for an amplifying device. On November 20, 1947 Brattain and Gibney wrote the disclosure which was witnessed the same day by J. Bardeen and H. Moore. This disclosure resulted in the first patents: *"Three-electrode circuit element utilizing semiconductor materials"* (2,254,034 filed on February 26, 1948) in a series of three patents covering the discovery of the transistor effect. In this disclosure they suggested that solids could be used instead of the electrolyte (Fig. 10.4).

On Friday, November 21, 1947 Bardeen presented to Brattain and Gibney a new idea for an amplifying device. He suggested insulating the point contact wire and inserting it into an electrolyte to contact the semiconductor surface. The bias of the plate electrode would control the surface electric field underneath the point contact and therefore control the current flowing into the sample (Fig. 10.4).

Gibney and Brattain started work immediately. Gibney replaced distilled water with glycol borate extracted from electrolytic capacitors and Brattain coated a point contact wire with insulating wax. After several unsuccessful attempts, mainly because the insulating wax did not completely insulate the very end of the point contact wire, the experimental setup worked. Brattain recorded in his notebook: *"p-type*

Fig. 10.2   Brattain notebook (November 17, 1947)

Fig. 10.3 Brattain and
Gibney's experimental setup
(November 20–21, 1947)

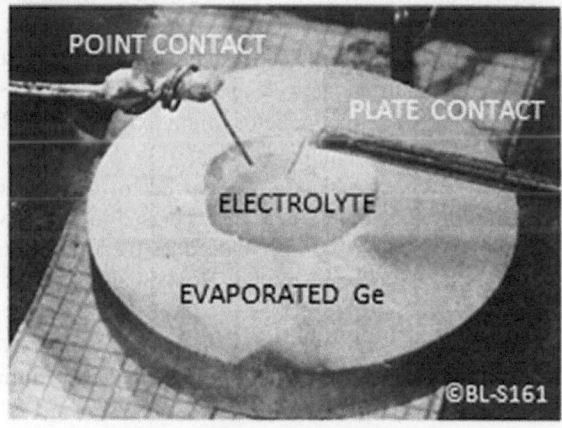

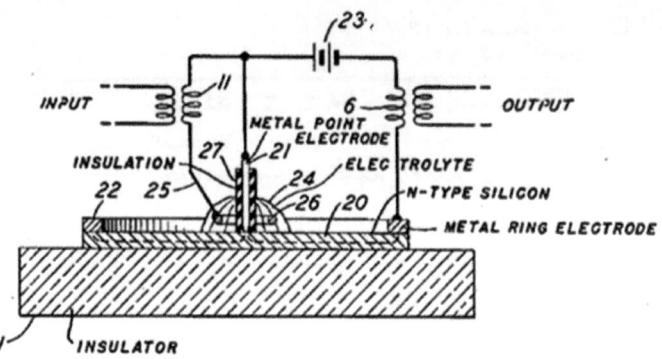

Fig. 10.4 Bardeen, Brattain, and Gibney's amplifying device (November 21, 1947)

germanium of thickness 500 Å and glycol borate as the electrolyte, a potential change of 2 V on input conductor resulting in a current range through the layer of 10^{-4} amperes."

Bardeen was not in the laboratory when Brattain and Gibney ran the successful test. When Brattain returned home, he called Bardeen and said[9]: *"We should tell Shockley what we did today"*, the idea being to call him to share the good result. But during the weekend Brattain changed his mind, and on Monday morning gave clear instructions to the members of the group: *"I swore them all to secrecy. They weren't supposed to know anything about this."* From now on, Shockley was provided only with selective results of Brattain's experimental work.

During the weekend, November 22 and 23, 1947, Bardeen wrote several pages in his notebook evaluating the data acquired by Brattain and Gibney. In these notes, he speculated about the presence and role of the surface layer which *"may be controlled by an external electric field."* He suggested that, if the magnitude of the electric field between the control electrode and the semiconductor surface was large enough, the opposite type of semiconductor might exist at the surface, creating the p-n (or n-p) boundary inside the sample (Fig. 10.5).

Based on the presence of the *"surface layer,"* Bardeen discussed the mechanism of current flow from a point contact biased in the forward direction and the current flowing into a point contact biased in the reverse direction. Bardeen's weekend work was witnessed on Monday, November 24, by W. Brattain, and on Wednesday, November 26, by W. Shockley.

Bardeen wrote a new disclosure titled *"Three-electrode circuit element utilizing semiconductor materials"* (2,254,033, filed on February 26, 1948)[10] covering the

---

[9]Brattain in an interview with A. Holden and W.J. King, AIP, June 1964.

[10]There is something unusual in Bardeen's patent. The semiconductor material is labeled in the figures as silicon, whereas the text describes the anodic oxidation of germanium introduced by Gibney.

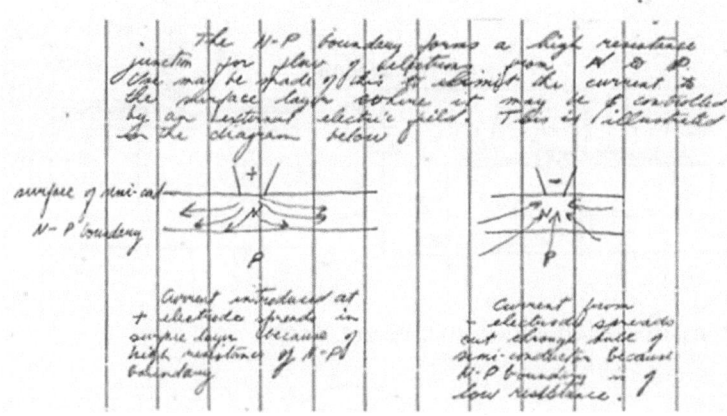

Fig. 10.5 Bardeen's theory of current flow from a point contact through a surface layer (November 22 and 23, 1947)

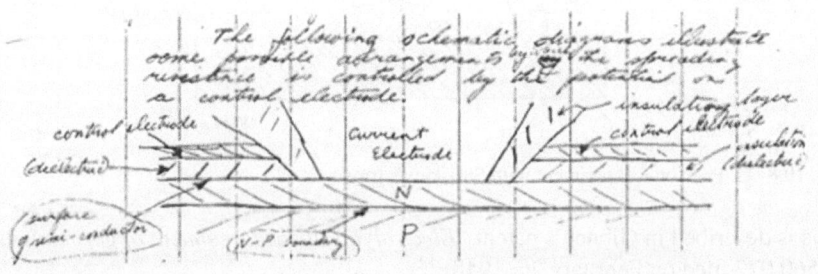

Fig. 10.6 Bardeen's disclosure of the patent 2,254,033, filed on February 26, 1948

control of the spreading resistance by the potential of the control electrode[11]. In the same disclosure Bardeen proposed to isolate the control electrode from the semiconductor surface by using an "*insulating dielectric*" instead of an electrolyte (Fig. 10.6).

In the same disclosure Bardeen proposed not to rely on the surface layer originating from the electrical field of a control electrode but rather to use a semiconductor sample with intentionally prepared "N-P boundary". The most important suggestion in Bardeen's disclosure is replacement of the electrolyte with polystyrene or oxide film (Fig. 10.7).

The new problem now was how to prepare the n- or p-surface layer. Gibney developed anodic oxidation treatment of n-type germanium using glycol borate.

---

[11]Brattain's and Bardeen's terminology is not consistent. Brattain used the terms "*reference electrode, ring, plate or upper capacitor plate,*" while Bardeen used the term "*control electrode*".

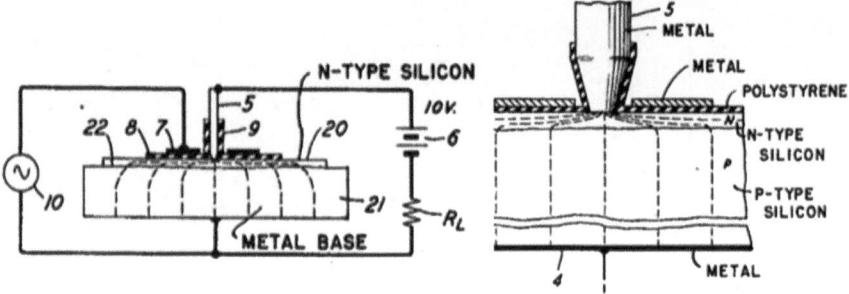

Fig. 10.7 Bardeen field-effect device (U.S. Patent 2,254,033)

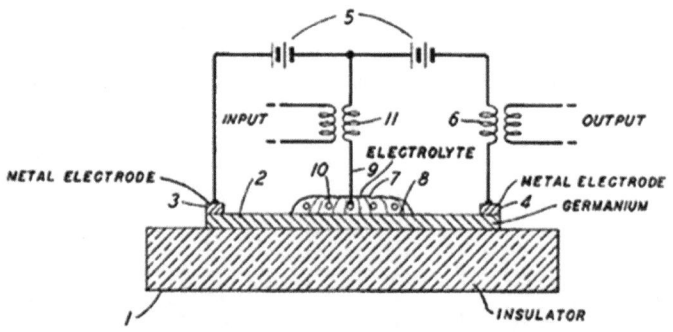

Fig. 10.8 G. Pearson's experiment with Shockley's triode

This is described in Gibney's patent *"Electrolytic surface treatment of germanium"* (2,560,072 filed on February 26, 1948).[12]

Shockley, in parallel, worked on the theoretical aspect of semiconductor phenomena: diffusion theory of minority carriers in a uniform p-layer and Zener diode. When he saw the result of Brattain and Gibney's experiment, he immediately proposed a new experiment. He suggested that a drop of electrolyte placed across a p-n junction operated at the high reverse bias might be used to obtain the voltage gain. Placing a drop of electrolyte at a localized p-n junction formed surface channels which increased the reverse current. G. Pearson verified Shockley's idea with a sample used for Shockley's investigations of reverse biased p-n junctions in an attempt to develop a new kind of lightning arrestor.

Encouraged by Brattain and Gibney's experiment, G. Pearson returned to the original Shockley triode proposal of 1945. In his notebook Pearson described Gibney's method of preparing a p-type germanium film evaporated onto a quartz plate with metal contacts on both ends of the sample and then he recorded data from electrical tests. By changing the biasing of the glycol borate electrolyte, he could reduce the resistance by 28.4% or increased it by 34% (Fig. 10.8).

---

[12]Gibney's patent uses the term *"transistor"* for the first time.

On Thursday, December 4, 1947 Pearson concluded: *"We then want to do the same curve on n-type which should decrease with plus voltage and increase with minus. This is a moral victory because it is a positive result on the field effect [for] which we have been looking for so long. It is the use of the Gu (glycol borate) which enables us to get the high fields necessary to successfully perform the experiments as well as using the crystalline films."*

Shockley noted that the experiments performed with the evaporated semiconductor thin films[13] the group was using did not exhibit normal rectifying characteristics. Encouraged by Pearson's accomplishment, Shockley, Pearson, Bardeen, and Brattain had a lunch meeting on Monday, December 8, 1947 and they discussed the future direction of work. After Bardeen's first, and Gibney's second key contributions, Shockley provided a third key suggestion: the collector must be biased to high reverse voltage. This suggestion resulted from Shockley's work on Zener diodes (Fig. 10.9).

The problem was that the group did not have materials capable of withstanding the high reverse voltage. For this reason, at this lunch, Bardeen and Brattain had only one choice: to use the Purdue germanium sample.

It is not clear why Bardeen's device, described in his disclosure dated December 15, 1947, still labeled all drawings with a reverse bias of −10V. Brattain's experimental data confirmed that, with such a low reverse bias, no transistor effect was observed, and in his subsequent experiments he used −90V.

The sample provided by Purdue University was referred to mysteriously as *"high back-voltage germanium"*. Today, the high back-voltage is called maximum reverse voltage and its magnitude depends on the purity (and hence the resistivity) of the germanium. Typically, within the resistivity range of 0.1–50 Ωcm the peak of the back-voltage ranges from 35 to 350 V. Such materials were needed for radar development at MIT Rad Lab. It turned out that the detector, in the form in which it was used in Britain and the U.S., was developed by the Germans before the war. In 1942, the NDRC had assigned germanium development to Purdue University and production of detectors to Sylvania and Bell Holmdel Laboratories.

Purdue had a duty to report progress every six months to Bell Holmdel Laboratories with a group from the armed services. Methods for melting and producing high-purity germanium ingots were perfected by S. Benzer and R. M. Whaley in mid-1944.

Following up on the lunch meeting, Robert Gibney prepared a new sample of Purdue germanium coated with approximately 3000 Å of germanium oxide prepared by anodic oxidation; then using a wax mask he sputtered four gold spots on top of

---

[13]Contrary to common folklore, in 1947 Shockley's group did not have good crystalline germanium or silicon samples. The many experiments were conducted with polycrystalline or evaporated films on quartz plates prepared by R. Gibney. The germanium research project started in Bell Laboratories after the invention of the transistor in February 1948. Gordon K Teal wrote: *"I repeatedly made germanium research program proposal, starting February 1948, and on through June, July and August of that year, and suggested preparing single crystals. J.A. Morton gave me financial support and encouragement."* [G.K. Teal, *"Roots of creative research"*, IDEA, Vol. 11 (1967) pp. 1–6].

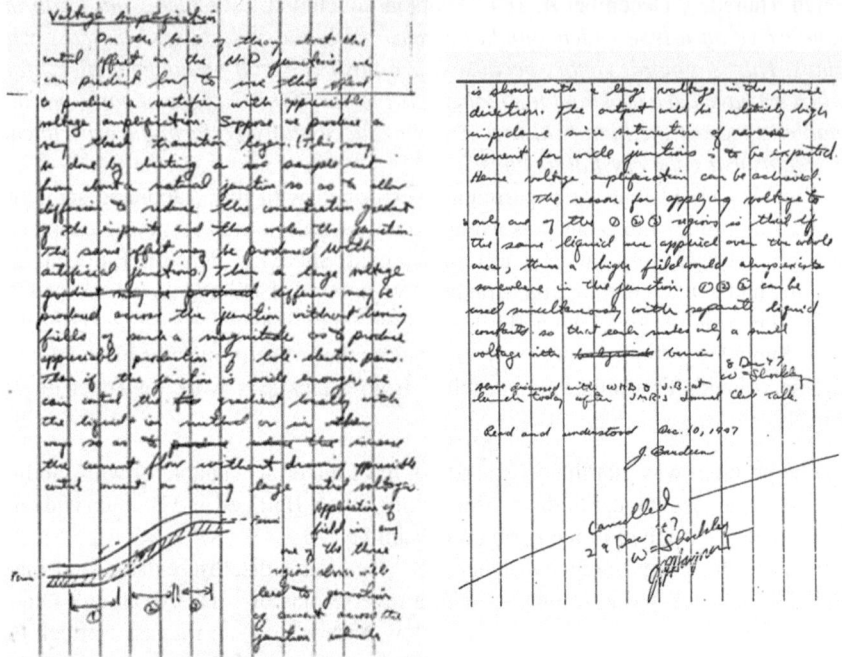

Fig. 10.9 Shockley's notebook dated December 8, 1947: *"p-n junction operated at high reverse bias might be used to obtain voltage gain"*

the oxide (one with diameter about 20 mm and three with diameter 10 mm). The intention was to verify Bardeen's idea of insulating the reference electrode (now called the control electrode) from the semiconductor.

Brattain's summary of the three experiments in his notebook (Fig. 10.10) is puzzling. The page is dated December 4, 1947. Note experiment III, where Brattain wrote: *"Two points close together the potential on one point to modulate the current flowing from the other point to the silicon."*

Brattain wrote: *"At a conference on Nov 22 or 29 it was decided."* Both dates are Saturdays. There is no note on those dates in either Shockley's or Bardeen's notebook. It is not clear how the idea of two-point contacts close together originated.

On the week beginning December 15, 1947 Robert Gibney left for an interview at Los Alamos Laboratories. His wife suffered with asthma and they wanted to move to New Mexico. Gibney accepted the Los Alamos offer to become the head of the metallurgy group, and left Bell Laboratories at the beginning of 1948.

Up to this point all of Brattain's and Pearson's experiments used electrolytes and worked at low frequencies only (less than 10 Hz) due to the long time constant of the electrolytes. On Tuesday, December 16, 1947, Brattain repeated the experiment

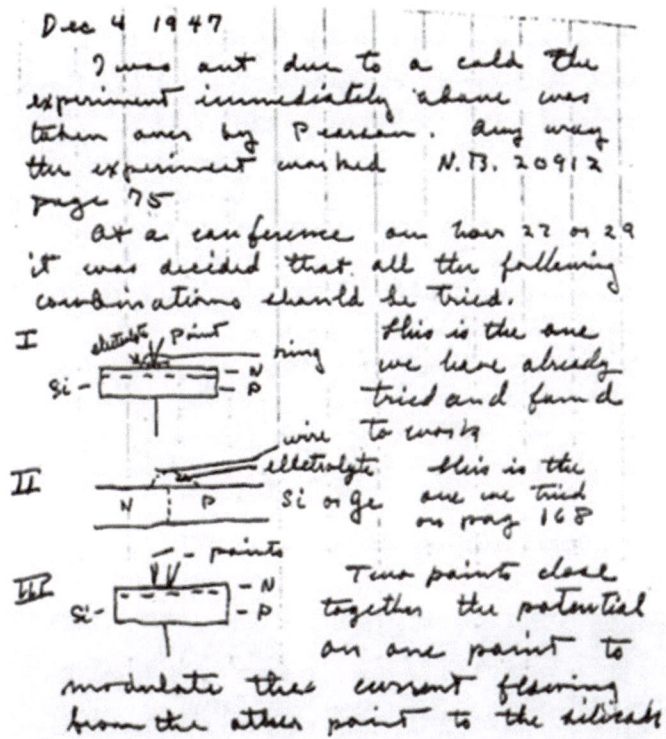

Fig. 10.10 Brattain's notebook, dated December 4, 1947

from November 21, but instead of the evaporated film he used a high back-voltage germanium sample prepared by Gibney in the previous week (Fig. 10.11).

When Brattain tested the sample, he found that germanium oxide was very porous and by accident he punched a hole in the large gold spots, creating a direct contact with the germanium. Inspired by experimental arrangement III in his notebook page for December 4, 1947, just out of curiosity, Brattain applied a positive bias to a gold spot and a negative potential to the point contact in the proximity of the gold spot edge. When the point contact was very close to the gold spot, he noticed a small voltage gain, by about a factor of 2, and relatively large power gain. The voltage gain was almost independent of frequency up to 10 kHz.

Brattain later described his experiment as follows[9]: "*We had a very specially chemically etched germanium surface on which the gold was evaporated. Gibney took it and did this and brought it back to me, and I started to investigate it. I inadvertently shorted the point to the gold film in the nice center hole so I got practically no data there. I was disgusted with myself of course, but decided there was no reason why I shouldn't go a round with the point on the edge of the gold to see if there was any effect, even if the gold was covering only part of the surface.*"

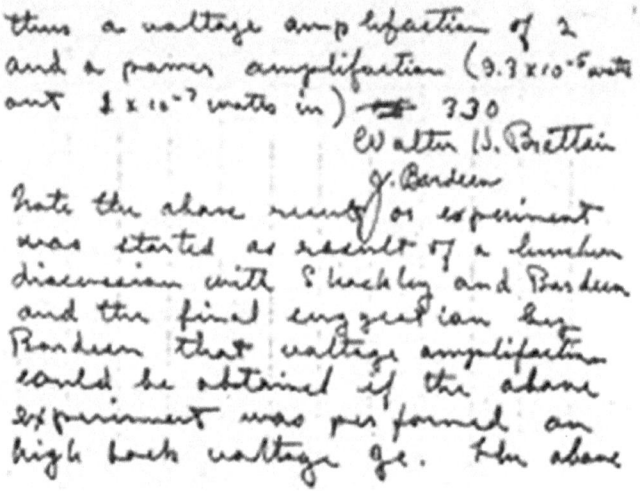

Fig. 10.11   Note in Brattain's notebook, dated December 16, 1947

Fig. 10.12   The very first experiment exhibiting the "transistor effect." (Brattain's experimental setup from December 16, 1947)

Unfortunately, neither Gibney nor Brattain recorded details of the "*special chemical etch.*" It is very difficult, if not impossible, to reconstruct Brattain's experiments, because he frequently changed terminology and almost never defined his method of measurement, the instrumentation used, or the sample of material used in their experiments. Brattain's samples were used over and over with different surface conditions. A typical example of Brattain's sample descriptions is shown in Fig. 10.14.

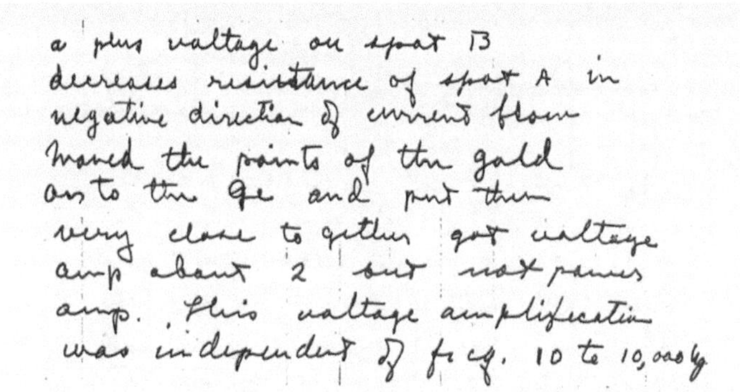

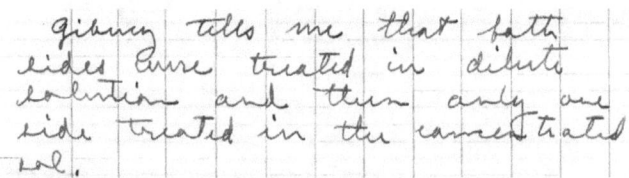

*a plus voltage on spot B
decreases resistance of spot A in
negative direction of current flow
moved the points of the gold
on to the Ge and put them
very close together got voltage
amp about 2 but not power
amp. This voltage amplification
was independent of freq. 10 to 10,000 ½*

Fig. 10.13   Brattain's notebook, dated December 16, 1947

*Gibney tells me that both
sides were treated in dilute
solution and then only one
side treated in the concentrated
sol.*

Fig. 10.14   Brattain's description provides no details of sample preparation

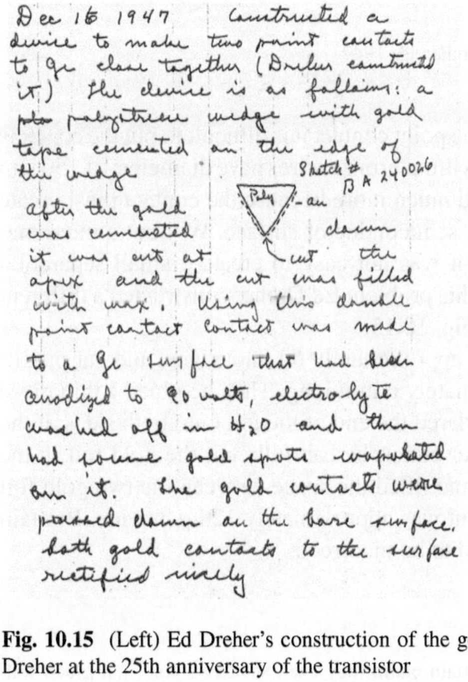

*Dec 16 1947   Constructed a
device to make two point contacts
to Ge close together (Dreher constructed
it.) The device is as follows: a
plus polystyrene wedge with gold
tape cemented on the edge of
the wedge*

*Sketch #
B A 24 0026*

*after the gold
was cemented
it was cut at
apex and the cut was filled
with wax. Using this double
point contact contact was made
to a Ge surface that had been
anodized to 90 volts, electrolyte
washed off in H₂O and then
had some gold spots evaporated
on it. The gold contacts were
pressed down on the bare surface.
both gold contacts to the surface
rectified nicely*

Fig. 10.15   (Left) Ed Dreher's construction of the gold contact device. (Right) Phil Foy and Ed Dreher at the 25th anniversary of the transistor

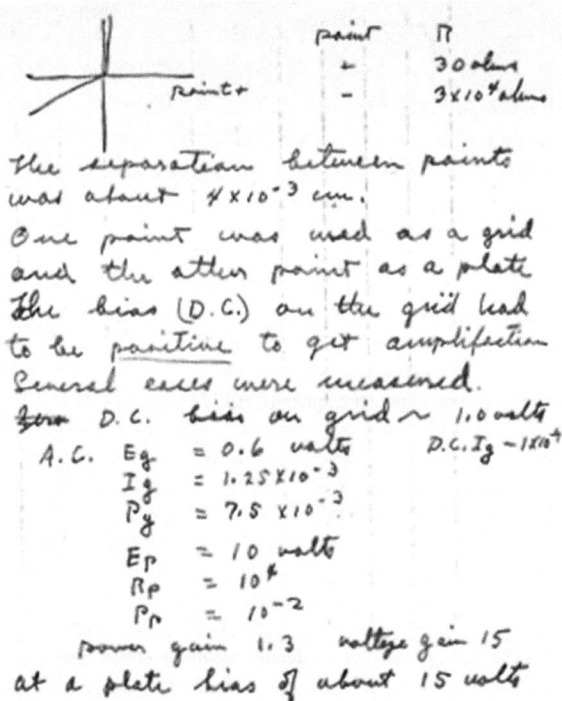

Fig. 10.16 Brattain's notebook, dated November 16, 1947

For experiments using more than one point contact it is difficult to put the contacts close together. Typical tungsten or beryllium bronze wires have diameters of 150 μm and these diameters cannot be reduced much more because the contact must create sufficient pressure to firmly touch the semiconductor surface. Without a micromanipulator (which was invented later) it was not easy to create a small separation between two point contacts. To solve this problem Ed Dreher constructed a different experimental arrangement, shown in Fig. 10.15.

Dreher used 8 mm plexiglass with approximately 60 mm edges, and cut part of a triangle, creating a flat part approximately 6 mm long. Then he glued a thin piece of gold foil[14] over the flat part and soldered the end of the foil onto both sides of the wire. With a razor mounted on a paper cutter, he carefully cut the gold foil in the middle of the flat part of the triangle and filled the space between the two gold foil parts with wax. The measured razor cut was approximately 120 μm wide. Brattain wrote 4×10^{-4} cm, but this was found to be incorrect.

---

[14]In later interviews both Bardeen and Brattain claim they used sputtered gold, not glued foil. Brattain in a later interview[8] claims that he etched the oxide before the gold evaporation. Gibney disputed such comments.

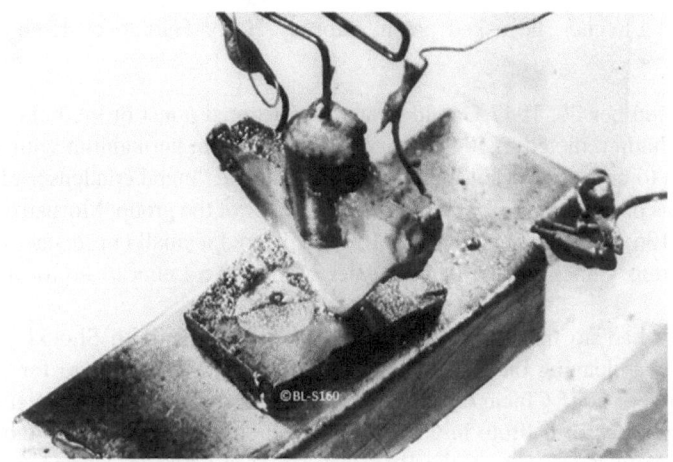

Fig. 10.17   Point contact "circuit element"

> The center point where Brattain shorted the gold spot is clearly visible in this image. This is
> a photograph of the actual Brattain and Dreher setup. The photographs used in the published
> literature show the replica built by Robert Mikulyak for the 25th anniversary of the invention
> of the transistor. To the author's knowledge this is the only available photograph of Dreher's
> actual experimental arrangement.

On December 17, 1947 Brattain used the same piece of germanium with gold spots
but applied a gold contact to the anodized surface in the space between the large and
small gold rings. When he checked the contacts between gold and semiconductor,
the gold foil did not create a good ohmic contact; the reverse current was relatively
large. Yet, by biasing one gold contact with $+2$ to $+3$ V and the other with -90 V,
the device exhibited voltage gain and power gain.

On the same day Shockley sent a memo titled "Concerning the Report on Semi-
Conductors" to Harvey Fletcher, and scheduled a demonstration for December 23,
1947.

The demonstration on December 23 worked in the same way as on December 17
but Fletcher was skeptical and insisted on a more sophisticated proof. He suggested
they use the amplifier with feedback as an oscillator. H. Moore agreed to modify the
circuitry and they scheduled a new demonstration for the next day. It worked again!

The "transistor effect" whose invention is now solely assigned to Bardeen and
Brattain was the result of hard work by the following individuals: J. Bardeen, W.
Brattain, E. Dreher, P. Foy, P. R. Gibney, H. R. Moore, G. L. Pearson, and W. Shockley.

Bown and Fletcher did not want to share details with M. Kelly until they were abso-
lutely certain that all experiments are repeatable. Kelly learned about the December
demonstration in the first week of January 1948 and after a short discussion with

Bown and Fletcher, he asked patent attorney Harry Hart to evaluate what was patentable.

On December 26, 1947 Gerald Pearson performed a test of the field effect that Shockley had predicted in 1945. Pearson used thin film germanium with an evaporated gold to constitute what Shockley labeled as the "metal condenser plate." The sample was prepared by Gibney and a new member of the group, Morgan Sparks. By applying 135 V on the gold plate, Pearson registered a small but repeatable change in the current. Shockley and Pearson later submitted a Letter to Editor in Physical Review.[15]

Pearson had no time to celebrate and discuss details with Shockley, because Shockley was leaving on December 28, 1947 to travel to Chicago for a meeting at the University of Chicago scheduled for the next day. With the next meetings in January at Case Institute in Cleveland and another in Pittsburgh, he decided to stay over New Year's Eve in the Bismarck Hotel in Chicago. During two days of the holiday Shockley created a description of a novel device "Voltage and Current Amplification using N-P junction". On January 2, 1948 Shockley sent an express mail to Stanley Morgan. Morgan received the letter on January 5 and witnessed it with Bardeen the same day. The pages were later glued to Shockley's notebook.

W. Shockley returned to New York on Friday, January 9, 1948. On Monday of the next week, he learned that Bardeen and Brattain alone prepared the patent disclosure *"Three-electrode circuit element utilizing semiconductor material"* (Patent 2,524.035 filed on June 17, 1948). Shockley was astonished that they did not include himself or another member of the group as co-inventor. Shockley talked to Bardeen who had nothing to say. Brattain, however, was more shameless and flagrant. When Shockley asked why other members of group had not been listed as co-inventors, Brattain shouted *"oh hell Shockley, because I did the work."*[16] Since then, Shockley, Bardeen, and Brattain's relationship turned sour, and Shockley never asked Brattain to do any additional experiments. Shockley knew that Bardeen and Brattain had only a vague understanding of their devices and that their interpretations of experimental results were not complete. The original point-contact transistor often behaved in true negative resistor fashion, but was never understood.

Historians describing "the magic month" of the transistor discovery missed one important detail, in all four patents covering the invention of the "Three-electrode circuit element." Bell Laboratories patent attorney Harry C. Hart specified the semiconductor material *"which is the subject matter of an application of J. H. Scaff and H. C. Theuerer, filed on December 29, 1945, Serial No. 638,351."* This application, titled *"Rectifier and method of making it"* covers the material used in radar detectors. Shockley's group did not have any samples of such material from Scaff and

---

[15]W. Shockley, G.L. Pearson, *"Modulation of conductance of thin films of semi-conductors by surface charges."* Phys. Rev. Vol. 74 (1948), pp. 232–233.

[16]Brattain later modified his formulation as[8] *"oh hell Shockley, there is enough glory for everyone."*

Fig. 10.18 I-V
characteristics of the
point-contact transistor
(2N21)

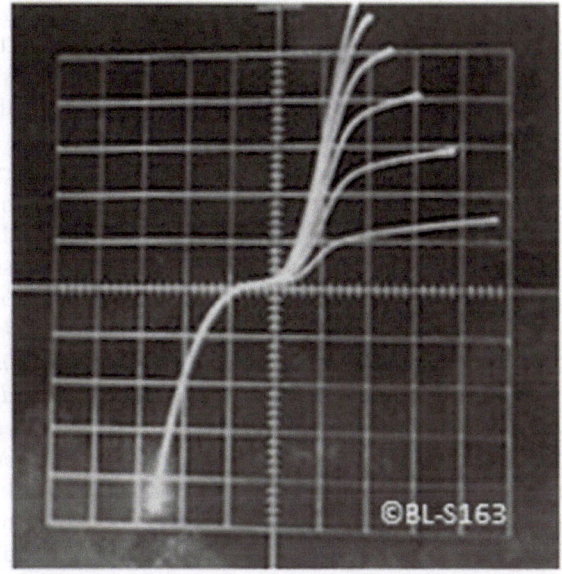

Theuerer's group, except some silicon polycrystalline samples. The Holmdel group
did not work on germanium during the war. In addition, Scaff and Theuerer worked
with Russel Ohl in the Bell Holmdel group, and Ohl had an unsolved feud with
Brattain and did not want to work with him. Hart was afraid to acknowledge that the
semiconductor materials used by Brattain in successful experiments originated from
Purdue University and were prepared by R. M. Whaley. Bell Laboratories did not
have their own germanium material until 1948.

It is not often recognized that Brattain's experimental setup of the discovery of
the transistor effect is a very different device than the commercial, so-called A-type
transistor, 2N21/22/23 which was designed by William G. Pfann, Leopoldo Valdes,
and Jack Morton, and manufactured by Western Electric. The 2N21 transistor was
manufactured with metallurgically prepared polycrystalline germanium which was
used in Western Electric's diode production.

Western Electric manufactured about 30,000 units. Typical I-V characteristics are
shown in Fig. 10.18. Note that the I-V curves of the point contact transistor do not
extend to zero for $V_C = 0$. Almost one volt is required to start conduction. This
means that switching in saturation will result in discontinuities of current.

After several years of development, the point contact transistor outlived its useful-
ness, and was considered completely obsolete. The point-contact transistor was
simply bypassed by advancing to other transistor types more easily manufactured
and with less manufacturing variances. Though rather crude and poorly understood
it can be regarded as the beginning of the era of semiconductor devices.

Shockley's sour relation with Brattain and Bardeen never recovered and the three
never worked together again. For a while, Bardeen and Brattain worked hard to

develop the theoretical basis of the transistor effect. Their papers[17,18,19] gave the first account of their theoretical ideas. However, many of the details of their work were wrong. Their ideas were amended later, but up to this day the underlying theory of Brattain's device is still imperfectly understood, even after three quarters of a century.

Brattain remained in Bell Laboratories until 1967. After the transistor effect invention, he did not participate in any important work. Brattain's colleagues remember him as being more than willing to argue for his viewpoint, and not welcoming suggestions about his work. He did not have a good understanding of semiconductor physics. Perhaps, one of the brilliant women in solid-state physics, Esther Conwell, provided the best characterization of Brattain[20]: *"Brattain was not nearly as sophisticated as some of the others, and good in the experiments that he did, but wouldn't make any real effort to interpret any of his data. When he found an effect with his point-contacts he went to Bardeen to work on the interpretation and suggest further experiments. And Brattain would try to stand up to Shockley, and it was just pathetic. Shockley would just wipe the floor with him. Shockley was a very, very bright man."* It is not surprising that Brattain is the only informant in Shockley's FBI file who provided a reference questioning Shockley's contribution to the transistor invention.

Bardeen was dissatisfied, firstly because Shockley did not include him in the work on the junction transistor, and secondly, because Shockley's publications describing the theory of junction transistors and minority carrier injection proved that Bardeen's transistor effect explanation[19] was wrong. Bardeen went over Shockley's head and complained to Ralph Bown. Bown called Shockley and warned him of the danger. Shockley suggested to Bardeen that he could work instead in Jack Morton's group on the point-contact transistor. Morton did not like Bardeen and did not want him in his group. With a decreasing number of options available, Bardeen wrote a demanding memo to Mervin Kelly on July 24, 1951 and requested a personal meeting. Kelly was well aware that the power behind the transistor invention was Shockley, he could see how brilliant Shockley was, and he liked Shockley's way of explaining things with simple language. In addition, Kelly was under increasing pressure to increase profit for the corporation, and he liked the changes Shockley had made in his group, moving from basic research toward practical device-oriented research. Kelly refused to meet Bardeen. Bardeen left Bell Laboratories angrily at the end of 1951 and joined the University of Illinois at Urbana-Champaign. It was atypical for Bell Laboratories that there was no farewell party for Bardeen.

---

[17] J. Bardeen, W.H. Brattain "The transistor, a semi-conductor triode", Phys. Rev. Vol. 74 (1948) pp. 230–231.

[18] W.H. Brattain, J. Bardeen, "Nature of the forward current in Germanium point contacts", Phys. Rev. Vol. 74 (1948), pp. 231–232.

[19] J. Bardeen and W.H. Brattain "Physical principles involved in transistor action", Phys. Rev. Vol. 75. (1949), pp. 1208–1225. Authors have given a discussion of the matter, but their treatment is restricted the special case in which all of the forward current is carried by holes.

[20] E. Conwell interview with B. Ashrafi, AIP, January 22, 2007.

By the end of 1947, Shockley's understanding of semiconductor physics was unmatched by anyone else. With his phenomenal physical insight and ability to consider only the essential factors of any technical matter, Shockley embarked on the most productive period of his life.

Chapter 11
The Junction Transistor

"Stubborn people are strong-willed people, it's not easy to live with them. Once you stop viewing this personality trait as negative, stop butting heads with them, and learn how to deal with a determined leader, you will become the top gun. Thank your stubborn person for making you a better person."

June Silny, "Ways to deal with stubborn people".

Bardeen and Brattain's failure to acknowledge the work done by others can be explained by human nature—pride, arrogance, ignorance, plain self-interest, or in this case, by jealousy that Shockley's ability to solve the problems of solid-sate physics was better and faster. Almost all post-renaissance era inventions are the result of the combined efforts of several contributors. Ruling out R. Gibney and W. Shockley from the invention of the point-contact transistor by Bardeen and Brattain was a foolish decision. Shockley's disappointment was immense[1]: *"My elation with the group's success was tempered by not being one of the inventors"*. Historians characterized the issue differently[2]: *"He (Shockley) had felt burned after Bardeen and Brattain borrowed an idea he'd proposed during their meeting at lunch on December 8 to take "a big forward step" toward their invention."*

Bardeen and Brattain's continued work on the surface states after December 1947 brought them to a dead end because they did not recognize that point-contact probes with the probe spacing much larger than the thickness of the surface layer pass the current almost entirely through the bulk of the semiconductor and the surface state current is not detected. They realized this error after Shockley disclosed his p-n junction theory.

Shockley's theory of the p-n junction triggered enormous excitement within Bell Telephone Laboratories. In the period of time between the invention of the transistor effect and a public announcement of the discovery in July 1948, Director of Research Ralph Bown organized a development team headed by Jack A. Morton. The development of the point-contact transistor resulted in very different device than Brattain and Bardeen's experimental setup with flat gold strips contacting the germanium

---

[1] W. Shockley, *"The invention of the Transistor"*, Proc. of Conf. on the Public need and the role of the inventors, Monterey, CA, June 11–14, 1973.

[2] M. Riordan, L. Hoddeson, *"Crystal Fire"*, W.W. Newton & Co. NY 1997.

© The Author(s), under exclusive license to Springer Nature Switzerland AG 2021   103
B. Lojek, *William Shockley: The Will to Think*, Springer Biographies,
https://doi.org/10.1007/978-3-030-65958-5_11

surface. After the announcement of the junction transistor in the middle of 1951, Morton's group was disbanded and the point-contact transistor became extinct.

As with many great men in history, Shockley's setback sparked a change in strategy: he said to M. Sparks "*if they do not want to do it with me, I will do it without them.*" His motivation was colossal. Shockley always wanted to interpret experimental data based upon a sound theoretical foundation. He did not approve of speculation in physics. In contrast to what happened with the point-contact transistor, which was a result of trial and error experiments, Shockley constructed the theory of the new device first. When Pearson did the experiment with an evaporated germanium p-n junction with glycol borate placed over the junction during the first week of December 1948, Shockley was baffled as to what mechanism was causing the amplification. Shockley realized that, if a high reverse voltage is applied to a rectifying p-n junction, the current increases precipitously with a further slight increase of the reverse voltage. His focus changed to: "*interest on the physical phenomena and not on device application. My interest was in using electrical measurements to obtain fundamental scientific information about the basic processes.*"

Shockley's group was extended by new members who had worked on radar research during the war: R. M. Ryder, J. R. Haynes, H. Suhl, F. S. Goucher, W. C. Westphal, R. M. Mikulyak, E. Buehler, W. J. Pitenpol and others.

On April 24, 1947 Shockley recorded in his notebook the analysis of electrons in a p-type semiconductor. He correctly considered the diffusion of minority carriers and the generation of carriers in the space charge layer which gives rise to large currents in a reverse-biased junction. However, in order to find the concentration of minority carriers, he used a simplified and non-realistic boundary condition which made it impossible to understand the influence of minority carrier injection.

In Shockley's New Year's Eve work he proposed a "*Voltage and current amplifying device using n-p junction.*" Without concern for how the device could be assembled, he assumed that the p-n-p structure could be constructed by using diffusion. The n-region created by diffusion from the n-type semiconductor (labeled B in Fig. 11.1) should have a higher n-type doping at the contact B and lower at the surface. By applying a negative voltage to B, the potential barrier for the hole current could be lowered, allowing holes to reach the reverse-biased collector A_2. Although the energy band diagram resembled the diagram of a junction transistor, Shockley still did not realize the possibility of holes injected into n-material becoming the minority carriers in the n-type material. This contrasts with work Shockley did in April 1947, where he considered the diffusion of minority carriers in a reverse-biased p-n junction.

The problem was that Shockley did not consider the structures as three-terminal devices. He was still searching for an explanation of how the point-contact transistor worked and to determine whether or not an inversion layer played an important role in the transistor effect.

On January 23, 1948 Shockley recorded a new disclosure "*High power large area semi-conductor valve.*" His assumption was that evaporated thin film layers of n-p-n semiconductors with ohmic contact to each layer would allow modification of the concentration of electrons in the n-type base rather than the potential of holes in the p-layer (inversion layer). Once he drew the band diagram, Shockley recognized that

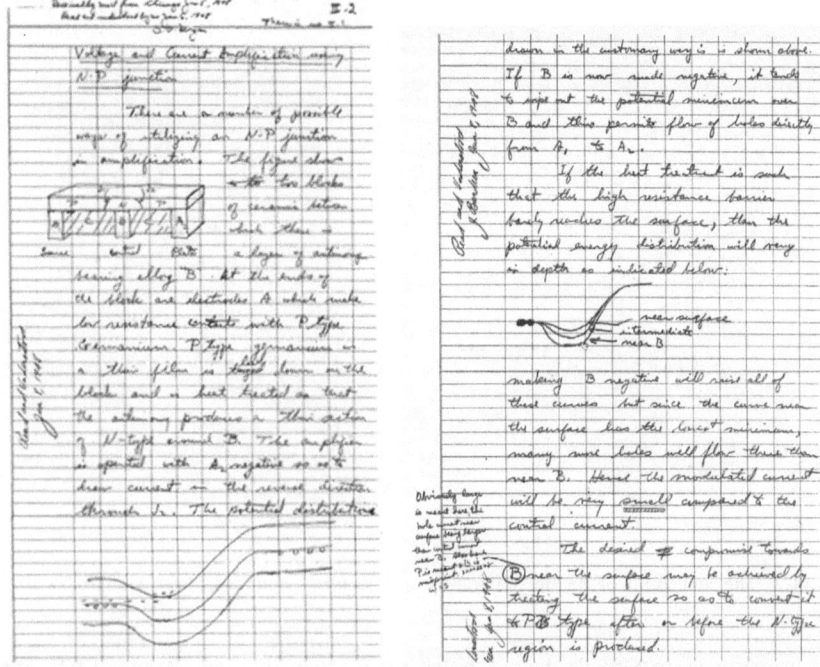

Fig. 11.1   Original concept of the junction transistor (Shockley's notebook, dated January 23, 1948)

the minority carriers diffusing through the middle layer would control the current of the reverse-biased collector. The disclosure was witnessed by J.R. Haynes on January 27, 1948. By drawing the band gap diagram (Fig. 11.2), Shockley realized that contacting the "middle" layer by an external voltage enabled the device to act as an amplifier.

Shockley's enthusiasm was bubbling, and his work habits were extraordinary: he worked from early morning to midnight, the weekends were no exception, and vacation was not in his vocabulary. He was focused only on his most important goal (and he also forgot that he had a family). He could not talk about anything else but p-n junctions.

Shockley told me about an encounter with Brattain when they met on the train after a January 1948 meeting of the American Physical Society in New York. Shockley enthusiastically explained his idea about the existence of an internal contact potential (ϕ_{bi}) in a p-n junction. Brattain consternated Shockley with his answer: instead of Brattain's typical answer "*it would not work*" he replied "*this is nonsense.*"

On February 13, 1948 John Shive recorded in his notebook results of his experiment with a thin slice of germanium of thickness 100 μm contacted with phosphor-bronze point contacts with diameter of 120 μm on opposite sides of the sample. Shive demonstrated good transistor action and put the Bardeen and Brattain explanation of transistor action in doubt. The current passed between emitter and collector over

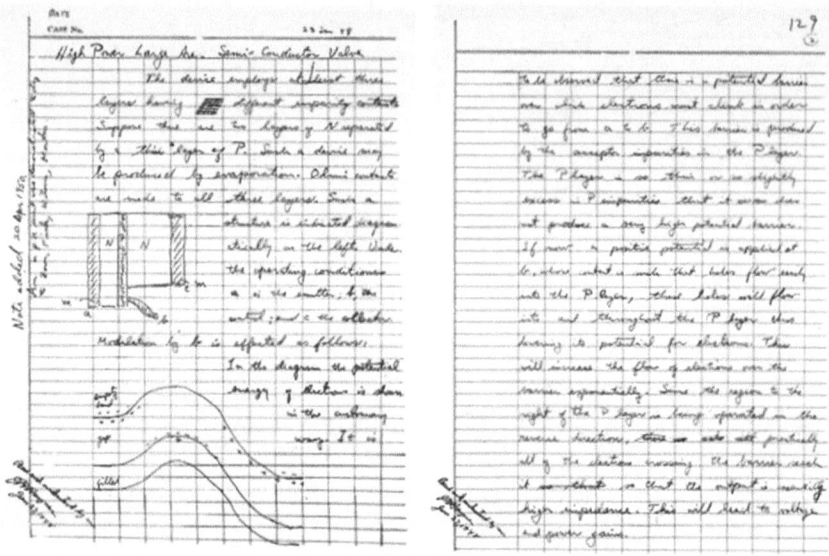

Fig. 11.2   The conception of the junction transistor (Shockley's notebook, dated January 23, 1948)

the surface was, in Shive's device, much longer than in a device where the point contacts had been on the same side of the germanium. If point contacts in Bardeen and Brattain's device were separated by the same distance as on the Shive device, the device would exhibit no gain. It was clear that while the point-contact transistor may have exhibited some surface effects, bulk propagation was surely also taking place and was probably the dominant effect.

Shive correctly assumed that holes would flow through the body of the sample. Although Shive introduced the term *"minority"* and *"majority"* carriers in his description, he did not include injection of minority carriers that diffuse through the body before they are collected by the reverse-biased collector. Shive reported the results of his experiment at a group meeting on February 18, 1948. After Shive's presentation, Shockley went to the blackboard and described his recently developed theory of the junction transistor and his theory of carrier injection by an emitter junction. Such current flow is also involved in the coaxial transistor of W. E. Kock and R. L. Wallace, Jr.[3]

That day, the concept of using a p-n junction rather than metal point contacts had been disclosed and the "junction transistor" was invented. The patent application for the junction transistor was filed on June 26, 1948 (U.S. patent 2,569,347). Shockley was paid $1.00 in advance by Bell Laboratories for signing the employment agreement. Prof. J. Gibbons described Shockley's 1948 patent on the bipolar

---

[3] W.E. Kock and R.L. Wallace, *"The coaxial transistor"*, Electrical Engineering, Vol. 68, (1949), pp. 202–203.

junction transistor as *"the most extraordinary use of imagination I have ever seen in a patent."* Bell Laboratories did not actually file the patent for six months, in part because *"Shockley wasn't sure he had a way to describe how to make it."* What is less known is that this patent included additional claims for heavy doping of ohmic contacts, heterojunctions with wide energy gaps to increase emitter efficiency, many layer structures, and transit time effects.

During the summer of 1948 Shockley turned his attention to the theory of the p-n junction. He summarized his ideas in a concise description of p-n junction theory in the form it is used today. He introduced the concept of the quasi-Fermi level (which he called "imref"), "internal contact potential" (now called "built-in potential ϕ_{bi}") and created a theory of depletion approximation. The work was published in a classic paper *"The theory of p-n junctions in semiconductors and p-n junction transistors"* published in the Bell System Technical Journal[4] in July 1949.

Shockley's detractors claim that he copied the work of Boris Davydov and that Bardeen was aware of minority carrier injection.

(1)   We can only speculate now how much of Shockley's work was influenced by Davydov's papers published between 1937 and 1939 before his excellent work was interrupted by the war. I found all three of Davydov's papers published in English in Shockley's library. The paper published by Davydov[5] in 1938 clearly spoke of *"the concentration of the free electrons in the "holes" semi-conductors and the concentration of "holes" in the normal electron semi-conductor. They must, therefore, be taken into account, no matter how small they be."*

The Rectifying Action of Semi-Conductors   93

$$j_s \approx 2kT \left(a' \mu_i' n_i' + a'' \mu_i'' n_i'' \right). \tag{25a}$$

In this expression for j_s only the lesser of the concentration $n_{i,a}^2$ entered for each of the semi-conductors (i. e. the concentration of the free electrons in the «holes» semi-conductor and the concentration of «holes» in the normal electron semi-conductor). They must, therefore, be taken into account, no matter how small they be.

Because Western historians adulate American technology and have a hard time to acknowledge Soviet progress, we seldom learn about this interlude. I witnessed a conversation between W. Shockley and a legend of Soviet semiconductor physics, Abram Ioffe (Fig. 11.3). I learned that Shockley highly appreciated work done at the "Physico-Technical Institute" in Leningrad.[6]

---

[4]W. Shockley, "The theory of p-n junctions in semiconductors and p-n junction transistors", BSTJ, Vol. 28 (1949, pp. 435–489. This article should be mandatory reading for everyone working in the field of semiconductor devices.

[5]B. Davydov, "The rectifying action in semi-conductors", J. Tech. Phys. USSR, Vol. 5 (1938), pp. 87–95. Davydov's papers were also available to Brattain and Bardeen, as we find from the references they cited.

[6]The institute is now called Физико-технический институт имени А.Ф.Иоффе Российской академии наук.

Fig. 11.3 Two legends of semiconductor physics: William Brandon Shockley and Abram Fedorovich Ioffe (1960)

(2)   Shockley supplemented Bardeen's transport of holes through "*the surface of semiconductor and N-P boundary*" (later called the inversion layer, or barrier layer) by the mechanism of minority carrier injection which provided the rationale for proceeding with the further development of the transistor.

Bardeen and Brattain refer to the injection of excess holes from the emitter into a p-type inversion layer at the surface, and the migration of these holes to the collector. This corresponds to injection of holes into the p-type material and should not be confused with the concept of injection of minority carriers into the bulk material.

By mid-1948 the concept and theory of the junction transistor was reasonably well evolved, but Shockley was still struggling with the issue of the p-type surface layer. Shockley stated "*I am not now certain whether a surface layer or injection into bulk was what made the first point contact transistor amplify.*" Shockley embarked on original research work which put the minority carrier injection in a transistor on a solid footing and left no doubt who was the originator.

In spite of intensified research efforts at Bell Laboratories, no one was paying attention to growing single crystals of germanium. Even Shockley underestimated the importance of the semiconductor substrates at that time. Shockley assumed that, for research, it was sufficient to have samples prepared by cutting large crystals from polycrystalline ingots solidified from melt. The problem which no one recognized at that time was that polycrystalline germanium exhibits variations in resistivity at randomly occurring grain boundaries. In addition, at that time no one paid attention to the crystallographic plane orientation.

During the war, Gordon Teal,[7] a member of the Chemical Department, suggested a method of growing semiconductor crystals, but Shockley and also Scaff and Theuerer in the Metallurgy group rejected Teal's suggestion. G. Pearson stated[8]: *"Metallurgists always thought that the physicists were hogging the glory."* Fortunately, Jack Morton trusted Gordon Teal. On October 1, 1948, Teal completed the construction of a two meter tall crystal growing setup. Teal did all his work almost "illegally", in parallel with his official job assignment of developing silicon carbide for varistors in telephone handsets. By December 1948, J. Morton legalized Teal's work and he provided the crystalline germanium to Shockley's group and later to the production of 2N22 and 2N23 point-contact transistors.

Shockley suggested a series of experiments to investigate the mobilities of electrons in germanium under a high electric field. On December 1, 1948 E. J. Ryder and W. Shockley submitted an article[9] to the Physical Review. The paper used the term *"injection"* for the first time and the authors discussed the evidence for the idea that holes are actually introduced into n-type germanium by the forward current of an emitter point.

In 1948, Haynes and Shockley published another paper[10] about how to measure experimentally the mobility, the recombination lifetime, and the diffusion coefficient of minority carriers. The conductivity of a semiconductor material is proportional to the mobility and carrier concentration, thus to optimize semiconductor devices, investigation of the mobility with respect to the doping of a material has great importance.

They repeated very sophisticated experiments performed during 1948 and 1949 by Haynes, Pearson, Suhl, and Ryder with Teal's single crystalline material, and confirmed that the minority carrier lifetime was 20–100 times greater than for polycrystalline germanium. Shockley's group was gradually gaining detailed insight into the behavior of both types of minority carriers, and a detailed understanding of the injection efficiency, mobility, diffusion, and lifetime. The data provided a clear indication that Shockley's junction transistor device would work, but the proof was still missing.

The first working junction transistor was constructed by Robert Mikulyak. He constructed a "dropping unit" which dropped molten p-type germanium onto an n-type germanium slab heated close to melting temperature. The solidified droplet was later polished to create a slab p-n junction structure. On April 7, 1949, Mikulyak

---

[7]Few people know that in 1952 Gordon Teal was the only private individual who purchased the license for Western Electric's transistor technology, because he was planning to enter the transistor business.

[8]G. Pearson in an interview with L. Hoddeson, AIP 1976.

[9]E. J. Ryder, W. Shockley, "Interpretation of dependence of resistivity of germanium on electric field", Phys. Rev. Vol. 75 (1949), p. 310.

[10]J. R. Haynes, W. Shockley, "The mobility and life of injected holes and electrons in Germanium", Phys. Rev. Vol. 81 (1951) pp. 835–843.
J. R. Haynes, W. Shockley, "Investigation of hole injection in transistor action", Phys. Rev. Vol. 76 (1949), pp. 691–692.

Fig. 11.4 Surface of
polycrystalline n-type
germanium after a cleaning
etch, as used in the
production of 2N21
transistors

250 μm

©BL-S155

Fig. 11.5 p-n junction
prepared with molten
germanium (R. Mikulyak's
notebook, dated April 7,
1949)

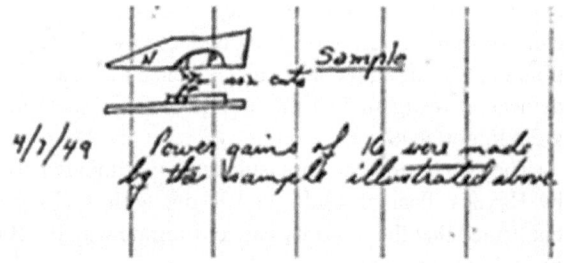

used the dropping unit to drop p-type melted germanium into the center of n-type
germanium, as shown in Fig. 11.5. The top of the p-type drop was ground and polished
until the p-region was about 180 μm thick. Then, using a vibrating saw, he cut two
grooves about 50 μm wide and attached tungsten wires to each groove, contacting
the n- and p-sections of the sample. The third aluminum contact was made to the end
of the sample.

Mikulyak's transistor exhibited a power gain of 16, but this structure was not
exactly what the claims in Shockley's patent specified:

> 1. A solid conductive device for controlling
> electrical energy that comprises a body of semi-
> conductive material having two zones of one con-
> ductivity type separated by a zone of the opposite
> conductivity type, said two zones being contigu-
> ous with opposite faces of said zone of opposite
> conductivity type, and means for making electri-
> cal connection to each zone.

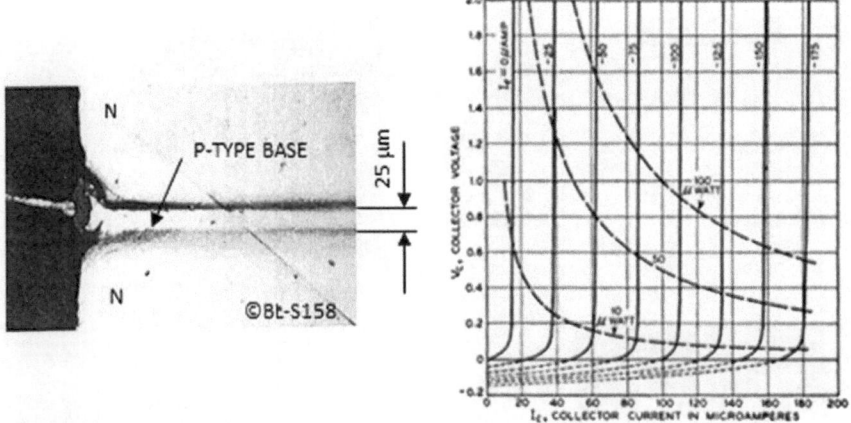

Fig. 11.6 The world's first NPN junction transistor (April 12, 1950)

However, this time good luck was on Shockley's side. During the summer of 1949 Gordon Teal discovered that a tiny pellet of antimony added to the growing n-type germanium caused the conversion of p-type germanium. In September 1949, after publication of Shockley's junction transistor paper, Teal suggested to Morgan Sparks that they make a junction transistor. Teal and John Little prepared the n-p-n germanium structure, and Morgan Sparks devised a tricky etch which raised the base layer so that it could be contacted by a gold wire. On April 12, 1950 Morgan Sparks recorded in his notebook the technique for making a base layer for contact in the fabrication of a junction transistor (Fig. 11.6).

On May 24, 1951 Shockley, Sparks, and Teal submitted to the Physical Review an article[11] which provided details of the construction and performance of the device. Mervin Kelly knew he had bet on the right horse, and invited Shockley for dinner. They recollected their first encounter in 1936 when Kelly foresaw the solid-state triode which Shockley, with his ability for getting the most out of the people who worked with him, had turned into a reality.

One of the great ironies of transistor history is that Shockley's junction transistor was not the field effect device he had envisioned in 1945, but rather the modern bipolar transistor, where the minority carriers injected into the bulk are crucial for the transistor action and the surface states play a less important role.

The very capable and clever R. Wallace, Jr., who had worked on the very early germanium junction transistors, built a machine to help automate the complicated manufacturing of the Sparks and Teal junction transistor. The most tedious processes with those transistors was bonding a thin gold wire to the narrow base layer. Wallace's machine started with little bars of germanium that contained the n-p-n structures,

---

[11] W. Shockley, M. Sparks, G.K. Teal, "p-n Junction Transistor", Phys. Rev. Vol. 83 (1951), pp. 151–162.

Fig. 11.7 Bell Laboratories "report of progress" (summer 1951 and 1952) with photographs of Teal, Sparks, and Wallace

made electrical contacts at the two ends, and used a thin gold wire to step automatically along the bar as an electrical probe until it found the voltage discontinuity at the p-n junction. It would then send a high current pulse through the gold wire, heating it until it alloyed with the germanium and made electrical contact with the base layer. His machine was called Mister Meticulous, because it meticulously probed along the semiconductor sample until it found a discontinuity at the p-n junction.

Bell Laboratories immediately launched a marketing campaign, popularizing the new transistor as *"running on as little as one-millionth of power of small vacuum tube"* (Fig. 11.7).

In less than one year, Shockley completed his Magnum Opus *"Electrons and holes in semiconductors"* on October 21, 1950. Shockley did not have a secretary. Every day, he wrote out a few pages by hand and brought them to R.M. Ryder for the first reading the next morning. The drawings were created by Bell Laboratories Drafting Department. When all was settled, Betty Sparks typed the manuscript and kept track of all changes.

The publisher D. Van Nostrand did not want to publish the book with the word *"holes"* in the title because the editor considered the title vulgar. After Bell Laboratories Director of Research Ralph Bown intervened, the book was issued in October 1950 with nine subsequent printings from 1950 to 1966 (Fig. 11.8). After more than seventy years, this book still makes enlightening reading today.

He dedicated the book to his mother with a note *"The encouragement and cooperation of Jean B. Shockley have been essential to the work."*

Bell Laboratories Series

**THIS COMPREHENSIVE BOOK ON TRANSISTORS
COVERS BOTH THEIR PROPERTIES AS CIRCUIT ELEMENTS
AND THEIR DESCRIPTIVE AND ANALYTICAL THEORY**

ELECTRONS AND HOLES
IN SEMICONDUCTORS

With Applications to Transistor Electronics

By WILLIAM SHOCKLEY

Member of the Technical Staff, Bell Telephone Laboratories
Author of Numerous Publications on Solid State Physics and Transistors

REFLECTING the point of view that led to the invention of the transistor, this book develops the subject from the broadest theoretical and practical standpoints. It shows how — with the advent of the transistor — the concept of the positive hole and that of its negative counterpart, the excess electron, have assumed new technological significance and a very much greater degree of operational reality, in Bridgman's sense of the word. From the theoretical viewpoint, the hole is an abstraction from a much more complex situation.

The achieving of the abstraction in a sound and logical way appears to involve inevitably rather detailed mathematical considerations. Experimentally, on the other hand, the behaviors of holes and electrons can now be inferred directly from the new post-transistor experiments with a degree of definiteness sufficient for most of the purposes of transistor electronics. This difference in level of abstraction between theory and experiment is reflected in the organization of this book. It provides in full the features needed to aid the engineer or student to master and apply this new science — including large numbers of clear-cut diagrams and graphs, and problems to show the orders of magnitudes involved, to supplement or extend the mathematics in the text, or to develop additional results.

You Can Judge — From the Following Brief Outline How Thoroughly This Book Introduces the New Subject of Transistor Electronics — From Basic Principles to Actual Use . . .   ©BL-S198

Fig. 11.8   D. Van Nostrand Company, Inc. Shockley book flyer (September 1950)

A perhaps unexpected reward for Shockley's work occurred on April 24, 1951. Shockley received a note from the National Academy of Sciences that he had been elected a member of the nation's most prestigious scientific institution. Shockley was one of the youngest people the Academy had ever rewarded with this honor.

Bell Laboratories significantly expanded the number of scientists working on transistor research at the beginning of the 1950s. Director of Research, J. Fisk announced[12] a new organizational structure with new departments: Shockley was appointed Director of Transistor Physics (Department 1450). Reporting to Shockley was J.A. Hornbeck, Department 1460 (Semiconductor Physics) and M. Sparks, Department 1470 (Transistor Feasibility). M. Sparks was also in charge of Department 1310 (Semiconductor Chemical Research). Many new scientists were involved

---

[12]Bell Laboratories Information Bulletin, Vol. 4, No. 42, December 3, 1953.

Fig. 11.9 Diffusion method
of forming a surface layer of
given conductivity

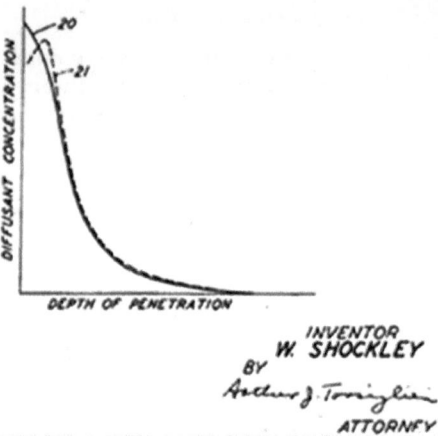

in the application of diffusion to the transistor technology inspired by several of Shockley's discoveries.

Shockley developed the application of impurity diffusion to form the semiconductor junction. The very first idea of high temperature diffusion was suggested in the junction transistor patent, filed on June 26, 1948: *"The electrode may be made of an antimony or phosphorus bearing alloy, such as a copper–antimony alloy or phosphor bronze so that heat treatment will cause antimony or phosphorus to diffuse into the p–type germanium changing it to n-type."* In patent application No. 496,201 "Method of forming large area p-n junction" submitted on March 23, 1953, Shockley proposed *"a method for forming surface layers of a given conductivity type and resistivity on semiconductor bodies to be used in semiconductor devices"* (Fig. 11.9).

Morris Tanenbaum who joined Bell Laboratories Chemical Physics Department in 1952, after graduation from Princeton University, worked on the chemistry and semiconducting properties of intermetallic compounds. In the fall of 1953, Tanenbaum was approached by Shockley and asked[13]: *"whether I would be interested in coming over to his group to see if we could make useful transistors out of silicon and determine if they were superior to germanium."*

There had been some attempts to make transistors using elemental silicon, which was known to have a higher energy gap and was naturally covered with a native thin layer of silicon dioxide. Those efforts had been unsuccessful, primarily because of what appeared to be the very low lifetime of minority electronic carriers. At that time the DuPont chemical company had become interested in silicon and they were making some fairly pure polycrystalline silicon. Tanenbaum worked with a technician Ernie Buehler, who was a craftsman in building apparatus and growing semiconductor crystals. (Buehler used radio frequency heating to melt materials with very high melting temperature in inert atmospheres without introducing impurities.)

---

[13]M. Tannenbaum in an interview with R. Colburn, IEEE 1999.

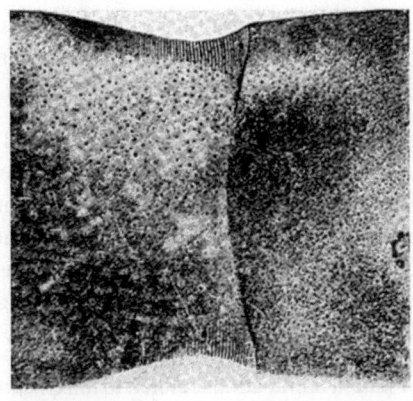 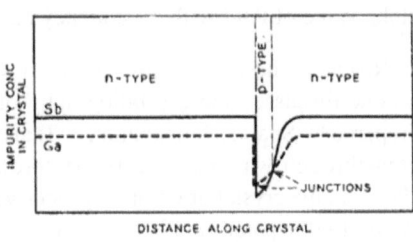

Fig. 11.10 The world's first silicon transistor (Bell Laboratories, January 1954)

Tanenbaum and Buehler started growing silicon crystals using high purity quartz crucibles containing the molten silicon. During the growth process, tiny amounts of doping elements (e.g., phosphorus, aluminum, etc.) were added to produce n-type and p-type regions to make n-p-n or p-n-p transistor structures. This was similar to the technique used to make the first Sparks and Teal germanium-grown junction transistors, which Buehler called "rate growing junctions." By January 1954, they could grow base layers with thicknesses of 12–15 μm and were able to make some n-p-n transistors with higher alpha values (a measure related to the amplification factor) (Fig. 11.10).

Alloying, even with good control of temperature and time, is not really a well-defined method for semiconductor applications. The junctions were irregular and their depths were not all under control.

Tanenbaum decided to publish the rate growing work and set it aside. The paper was presented at the IRE National Conference on Airborne Electronics in Dayton, Ohio on May 10, 1954. After Tanenbaum's paper, Gordon Teal, who was now with Texas Instruments stood up and said, *"We've made silicon transistors at Texas Instruments also. I have a handful of transistors in my pocket."* Tanenbaum asked Teal in discussion to tell him more about TI devices, but Teal replied that he was not permitted to talk about their work.

Authors[2] attribute the demonstration of the first silicon transistor to Texas Instruments, with Teal's demonstration of his first silicon transistor to the management of Texas Instruments being in April of 1954. However, Tanenbaum and Buhler demonstrated the silicon transistor in Bell Laboratories in January 1954. The technical details of the Texas Instruments growing technique were never made public.

Alloying was not sufficient to form the thin base layers needed for high-frequency operation. For this reason, Shockley's group decided to pursue an altogether different process for making transistor structures with much greater promise, namely diffusion techniques. The diffusion process can be better controlled than alloying and the main reason behind this decision was indeed better process control. On March 17, 1955 M.

Tanenbaum, D. E. Thomas, P. W. Foy. L. Valdes, and G. Kaminski fabricated the diffused emitter and base silicon transistor with a base thickness of 3.8 μm and an alpha value of 0.97 with a frequency cutoff of over 100 MHz.

Readers who have followed the above discussion should note that, up to 1954, no one discussed the crystallography of germanium or silicon. The p-n junctions prepared by Sparks, Tanenbaum, or Teal had a random 'dishing' shape. The important breakthrough was provided by Motorola engineer W.E. Taylor, who realized that choosing the crystal orientation properly made it possible to create parallel junctions (W.E. Taylor patent application *"Semiconductor assembly and method of forming same"*, Serial No 409,339 filed on February 10, 1954).

From 1952 to 1955 Shockley submitted 11 patent applications related to the circuit applications of semiconductor devices. In spite of his extreme work load, the patriot William Shockley still had time for frequent trips to Washington to consult with the Pentagon when the Korean war escalated. During the spring semester of 1954 Shockley was a visiting scientist at Caltech and returned to Bell Laboratories in July 1954. As soon as he returned to Bell Laboratories, he made a new important suggestion. Charles A. Lee, who joined Bell Laboratories in 1953 after graduating from Columbia University and became engaged in research on solid-state devices wrote[14]: *"the diffusion of impurities into germanium and silicon prompted Shockley's suggestion that the dimensional control inherent in these processes could be utilized to make a high-frequency transistor."* Together with P. Foy, W. Wiegmann, D. E. Thomas, and J. Klein, he constructed the first germanium diffused mesa transistor. The arsenic-diffused base had a junction depth of 1.5 μm. They created an emitter region by alloying aluminum to the depth of 0.5 μm. The resulting base thickness was about 1 μm. This first diffused germanium p-n-p transistor had a cutoff frequency of 500 MHz. The structure behaved in all respects according to Shockley's theory.

In patent application No. 392,971 filed on November 18, 1953, Shockley proposed diffusion from a solid source by heating with an electron beam. The diffusion from the solid source was a common technique used in the early days of semiconductor manufacturing. In patent application No 496,202 titled *"Semiconductor Device"* submitted on March 23, 1955, G.C. Dacey. C.A. Lee, and W. Shockley described the construction of the mesa type junction transistor with a concentration profile similar to that used in today's bipolar transistor (Fig. 11.11).

In parallel with work on diffusion, Shockley still pushed his original dream of the field effect transistor which he proposed on December 8, 1947 (Fig. 11.12). The work resulted in three patents (#2,744,970; 2,764,642; and 2,778,885). Shockley correctly analyzed the JFET before reduction to practice, just as he had done previously with the junction transistor.

Because of germanium's small band gap, germanium junction transistors have a large leakage current. J. J. Ebers was investigating methods for reducing the leakage

---

[14]C.A. Lee, presentation at the Semiconductor Device Conference of the IRE, Philadelphia, June 1955.

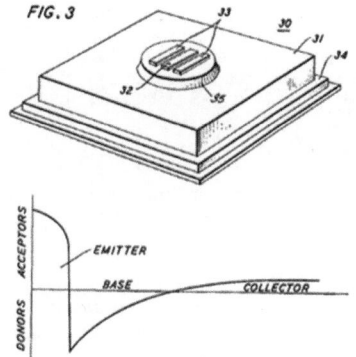

Accordingly, it is another object of the invention to facilitate the fabrication of a junction transistor whose base zone is characterized by a gradient in the concentration of the significant impurity atoms, which results in reduced transit times for injected minority carriers.

Fig. 11.11   The mesa transistor

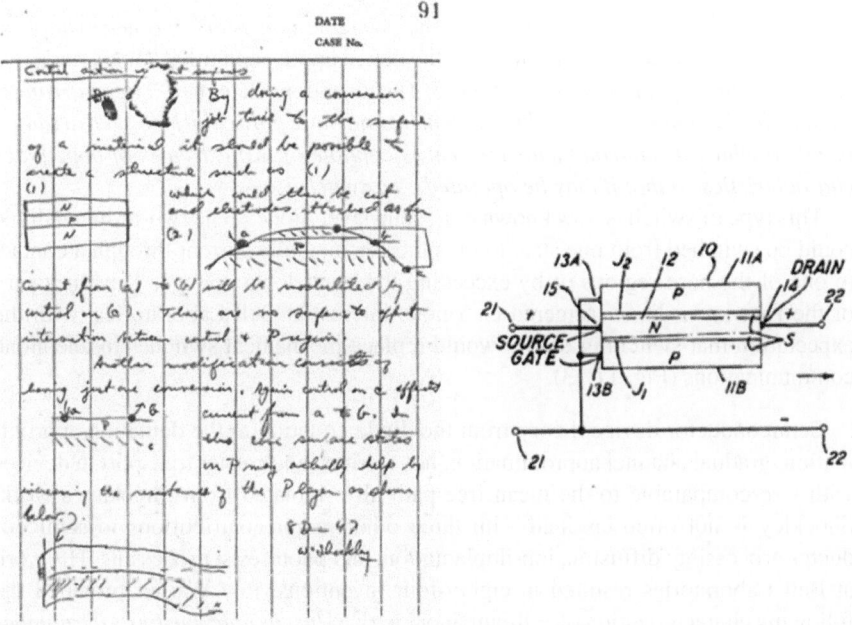

Fig. 11.12   Invention of the unipolar junction transistor (JFET) (December 8, 1947)

current. Using a hook-collector device previously proposed by Shockley, he discovered that a p-n-p-n structure can replace the combination of a p-n-p transistor and n-p-n transistor. On July 22, 1952 Ebers submitted a patent disclosure Serial # 300,235 titled "*Semiconductor signal translating device.*"

By analyzing Ebers' device, Shockley discovered that the structure could operate as a two-terminal device with bistable characteristics (US Patent 2,655,609 submitted

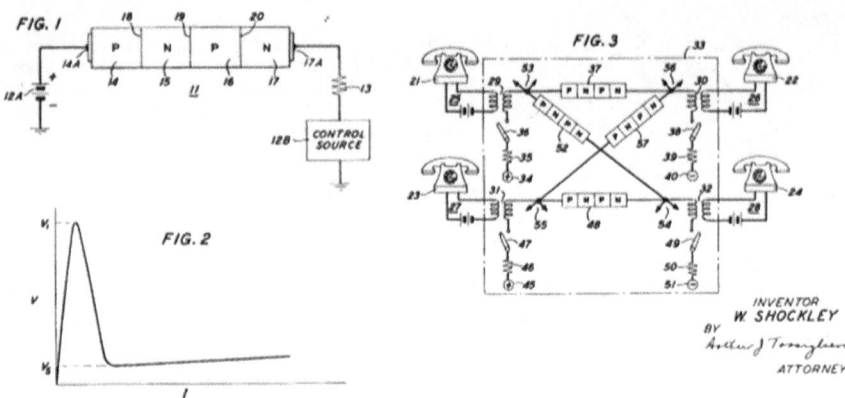

Fig. 11.13   The p-n-p-n four-layer diode (U.S. Patent 2,855,524)

on July 22, 1952). Theoretical work on the "bistable transistor" and negative resistance phenomena led to the invention of the patent application titled "*Semiconductor switch*," submitted on November 22, 1955. The invention relates to a "*circuit arrangement which includes a semiconductor element and more particularly to such arrangements in which a semiconductor element is capable of two extremes of impedance characteristics so that it may be operated as a switch.*"

This type of switch is now known as a "four-layer diode." The two-terminal diode could be switched from one state to the other by injecting current through a contact to one of the base regions or by exceeding the breakdown voltage. This discovery of the four-layer diode triggered enormous interest at Bell Laboratories, with the expectation that switching diodes would replace mechanical switches in telephone communications (Fig. 11.13).

Semiconductor device theory from the diode equation to the depletion approximation, gradual channel approximation, hot carriers, and carrier transport in devices with size comparable to the mean free path all originated from Shockley's work. Shockley is not often credited with three other major contributions to semiconductor processing: diffusion, ion-implantation, and photoresist processing. His work at Bell Laboratories resulted in eighty-four inventions. Ray Warner provided the following characterization of William Shockley[15]: "*Just as noteworthy as the number of Shockley's seminal ideas is the fact that these contributions ranged through realms of invention, engineering and science, and he was a star of all three. Shockley taught in his "Scientific Creativity" course that the essence of engineering is knowing what variables can be ignored. Shockley was a master of the simplifying assumption that got him to an analytical result. He had unusual skill in the frequent oral communications, delivering talks that were marvels of clarity.*" Former Stanford Professor, Jim Meindl said: "*When I was working at Stanford, I was fortunate enough to have an office three rooms away from Dr. William Shockley. His favorite quote was "Try*

---

[15]R. Warner in an e-mail to the author dated January 29, 2002.

Fig. 11.14 Shockley with his right-side steering wheel Jaguar

Simplest Cases" first, meaning to try simple things first when you explore new ideas, instead of getting lost in complications of the phenomenon you are exploring."

In 1953 Shockley was awarded the first Oliver E. Buckley Condensed Matter Physics Prize. The prize consisted of $20,000 and was established by the American Physical Society to recognize outstanding theoretical or experimental contributions to condensed matter physics. The citation reads: *"For contribution to the physics of semiconductors."*

Shockley enjoyed both the technical and aesthetic beauty of cars. His motto was: *"Any bad day can be fixed by driving!"* Shockley could not be happier than when driving. He was at the peak of his creative career and he had a new plan for the future (Fig. 11.14).

It is utterly bewildering why historians of science portray Shockley in Bell Laboratories as worse than a villain, while Brattain and Bardeen are only ever portrayed in a positive way.

Except for the experimental discovery of the "transistor effect," Bardeen and Brattain did not contribute any important work to semiconductor physics. Surface physics evolved into a separate and important solid-state physics discipline, but Bardeen did not continue working in this field. After 1950 he published many theoretical papers investigating various issues of superconductivity and this led to his second Nobel prize.

There is a significant difference between the discovery of the transistor effect and the invention of the junction transistor. The point-contact transistor never quite became a practical device; it was the junction transistor which triggered the microelectronics revolution.

Chapter 12
Shockley Semiconductor Laboratory

> *"Der Sieger wird immer der Richter und der Besiegte stets der Angeklagte sein"* (The victor will always be the judge, and the vanquished the accused.)
>
> Hermann Göring at Nuremberg Trials 1945–1946.
>
> (Military Legal Resources, U.S. Federal Research Division of Library of Congress)

Many years ago, I attended a three-day seminar organized by the young T.J. Rodgers at Stanford University. The topic on day one was how to answer a question "no." On the second day it was how to answer a question "yes," and on the last day how to answer a question "I do not know." Before attending the seminar, I already knew that some people talk for a long time without communicating any usable information. The seminar taught me a very effective way to get to the root cause of any problem.

All the protagonists of Shockley's Semiconductor Laboratory saga have already told their stories in numerous oral interviews and published texts. I noticed serious differences in accounts of the same event in published interviews, often contradicting the available historical documents. By applying Rodgers' methodology, that is, by asking for specific answers, I discovered that certain important facts and explanations concerning Shockley's Laboratory are omitted or never discussed, and those that are, when scrutinized, do not reflect reality.

In March 1954, Jack Morton and Shockley discussed the future of the transistor over drinks. Jack shared with Bill the advances in silicon float-zone refining methods and the progress in diffusion achieved by C.S. Fuller, M. Tanenbaum, and others, and suggested to Shockley that he start his own company. The progress in understanding the transistor and the enormous commercial potential of the four-layer diode set Shockley thinking that he should consider his own business activities. He had two

---

A large part of this chapter is based on Shockley's notebook #28 (author's archive).

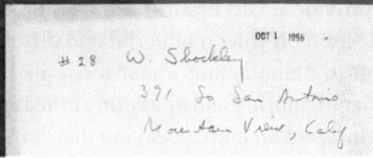

ideas which looked as though they could be developed into products: the unipolar JFET transistor and the four-layer diode. He had some new patentable ideas under his hat for these products and there was, in the foreseeable future, no competitor.

Shockley and two of his cohorts, M. Sparks and R. Wallace, Jr., started the initial exploration. As expected, Shockley first approached Mervin Kelly. Kelly contacted the third-generation member of the Rockefeller family, Laurence S. Rockefeller, who served as trustee of the Rockefeller Fund, and arranged a meeting. Rockefeller offered half a million but this was not enough to start a high-tech company. Similar negotiation with RCA and Raytheon also failed. Shockley actually started to work for Raytheon, but the company's management backed off from the original intention, and Shockley separated from the company after one month. The problem was that Shockley was banking on his name and his reputation, but he did not have any business plan, nor even a budget.

The Los Angeles Chamber of Commerce, in February 1955, cited W. Shockley and the valiant opponent of communism and adamant admirer of extraordinary ladies, Lee de Forest, for their ground breaking contributions to electronics. Arnold Beckman, who was the founder and president of Beckman Instruments, Inc., was also vice president of the Chamber of Commerce. Beckman also arranged travel for May to attend the lavish evening banquet, and organized a tour around the facilities of Beckman Instruments in Fullerton.

Beckman Instruments, Inc. evolved from Beckman's consulting work and was founded in 1935 in a garage in Pasadena, California. Beckman, a jazz musician and Caltech chemistry graduate, entered the laboratory market with the first commercially successful pH meter. The spectrometer and helipot potentiometer products were added at the beginning of the 1940's. A series of acquisition in the 1950's expanded the business with new sites in Europe.

Beckman Instruments went public in 1952, but the company was stagnating due to problems with the poor reliability of vacuum-tube-based instruments, which were not working very well. It was W. F. Gunning, from Beckman System Division, who suggested that Beckman should enter the semiconductor business and replace the vacuum tube electronics with semiconductors.

Beckman called Shockley on July 29, 1955 and invited him to Los Angeles. During the week-long discussion at the Balboa Bay Club in Newport Beach they agreed on the intention to start a new business. At that time Beckman Instruments, Inc. consisted of six divisions (Scientific, Helipot, Berkeley, Process, System, Spinco) and the Liston Backer plant. Beckman promised that 10% of sales would be assigned to R&D expenditure (which was $2.3 million in 1955). Beckman and Shockley agreed on a 3-year term, with $500,000 the first year and $250,000 each consecutive year. Shockley's vision was to create a new organization to provide a successful basic and applied research company making advanced devices based on silicon material and diffusion technology. He justified why he did not want to manufacture transistors—in 1955 there were already 29 U.S. based companies manufacturing and marketing transistors. Shockley did not want to be just a new player in a proven market, saying that "*other's had trampled already there*". He foresaw the biggest commercial success in field-effect transistors and a four-layer diode whose development would be supported by

Bell Laboratories and by the military. Shockley remained loyal to Bell Laboratories; he always wanted to use his unique knowledge of solid-state physics to create new devices. In 1955 the four-layer diode had the biggest chance of disrupting the existing mechanical switches used in telephone exchanges systems. Shockley's objective was to replace all 2 billion contacts in the Bell Telephone System with a cross-point solid state switch using the four-layer diodes.

In the early stages, Beckman wanted to have the new company in Fullerton and Shockley agreed. Beckman's attorney drafted a three and half page letter of intent. Shockley had 10 days to accept or refuse the offer. The letter of intent did not contain any specific information about time schedule, deliverables, manufacturing equipment, or type of product. The legal language prepared by L.N. Duryea, the counsel and assistant secretary of A. Beckman, was *"...important factors are suitable physical facilities, capable and congenial associates, a position of prestige and authority, with adequate voice in policy determination, and financial reward commensurate with performance which embodies, in adding to salary, some means for obtaining capital gains benefits."* It was so vague that even Shockley was confused and after hiring his attorney he requested some minor changes which Beckman accepted.

Shockley returned to Washington D.C. and Bell Laboratories. At Bell, Shockley negotiated with Jack Morton and Mervin Kelly for the future collaboration between the new Beckman Instrument Division and Bell Laboratories. The four-layer diode scored at the top of the priority list. Shockley returned to Fullerton and on September 1, 1955 in a one-man office started operation of Shockley Semiconductor Laboratories, Division of Beckman Instrument, Inc. A. Beckman acquired Bell Laboratories transistor license at the end of 1955. Morgan Sparks and R. Wallace, Jr. visited Fullerton on September 30, 1955 and met with Arnold Beckman to negotiate employment in the new division.

In the meantime, May leaked news to her neighbor, Stanford Prevost F. Terman. Terman immediately contacted Shockley and Beckman (Fig. 12.1). Terman argued with Beckman about the location of Shockley's laboratory. It is not clear why and how F. Terman and Professor John G. Linvill convinced A. Beckman to change his decision and move Shockley Laboratories to Palo Alto after their meeting in November 1955.

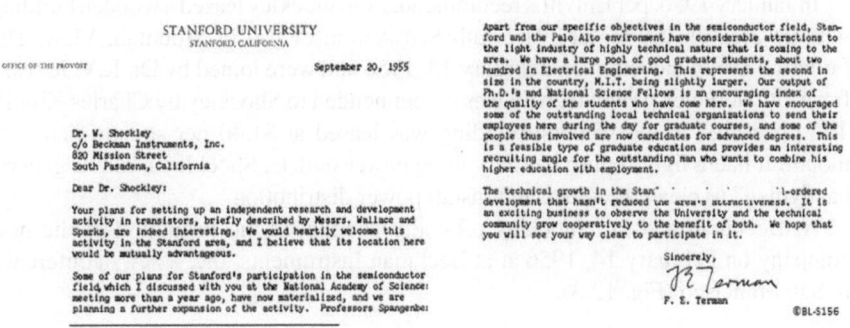

Fig. 12.1 F. Terman's reply to Shockley's letter dated September 20, 1955

Terman offered Beckman, who had recently acquired the Specialized Instruments Corporation (Spinco) located at Stanford University Business Park, full university support and possible expansion of the facilities (Terman offered a 10-acre lot at 1801 Page Mill Rd.). Linvill, before he joined Stanford University, was an assistant professor of electrical engineering at MIT, and he knew Shockley from his sabbatical at Bell Laboratories. Linvill suggested leasing temporary space for the Shockley Laboratory and moving all operations to the new building after its completion (Fig. 12.2).

On October 5, 1955 Shockley sent a letter to A. Backman:

> Another matter which will need serious consideration is whether or not we should start in Fullerton or in Palo Alto. At present, I think the decision may be postponed, but it may soon become important if we get strong reactions from future candidates such as Tanenbaum or some of the people I may try to find at Bell or RCA.

Fig. 12.2 W. Shockley's letter to A. Backman (October 5, 1955)

While still in Fullerton, Shockley hired three men with whom he had previously worked: Dr. Leopoldo B. Valdes (Pacific Semiconductor, Inc., and Bell Laboratories), Dr. G. Smooth Horsley (Motorola), and Dr. William W. Happ (Raytheon and Sylvania). Valdes was employee number 2 and he started at the Fullerton plant on October 10, 1955.

In October 1955 Shockley personally contacted his friends in academia and asked for good students of physics and chemistry. Shockley and Valdes *"made a new list of candidates whom we may approach at B.T.L."* and placed an advertisement in the February and March 1956 issues of Chemical and Engineering News. The location of the Shockley Laboratory was described as follows: *"The Laboratory will be located on the Stanford University Estate a short drive south from San Francisco."* These recruiting efforts resulted in twenty-five replies, none of them from a senior applicant.

In January 1956, per Linvill's recommendation Shockley leased a wooden building with an area of 465 m^2 at 391 South San Antonio Road in Mountain View. The Fullerton group moved in on February 13, 1956 and were joined by Dr. R.V. Jones, a fresh UC Berkeley graduate, who was recommended to Shockley by Charles Kittel[1]. The open area in the wooden building was leased at $1.40 per square meter/per month. It had a light fixture but only three power outlets. Shockley hired "a general handyman" to clean the space and install power distribution.

A. Beckman and W. Shockley officially announced the formation of the new company on February 14, 1956 at at Beckman Instruments, Inc. a news conference in San Francisco (Fig. 12.3).

---

[1] C. Kittel nominated Bardeen, Brattain, and Shockley to the Nobel committee.

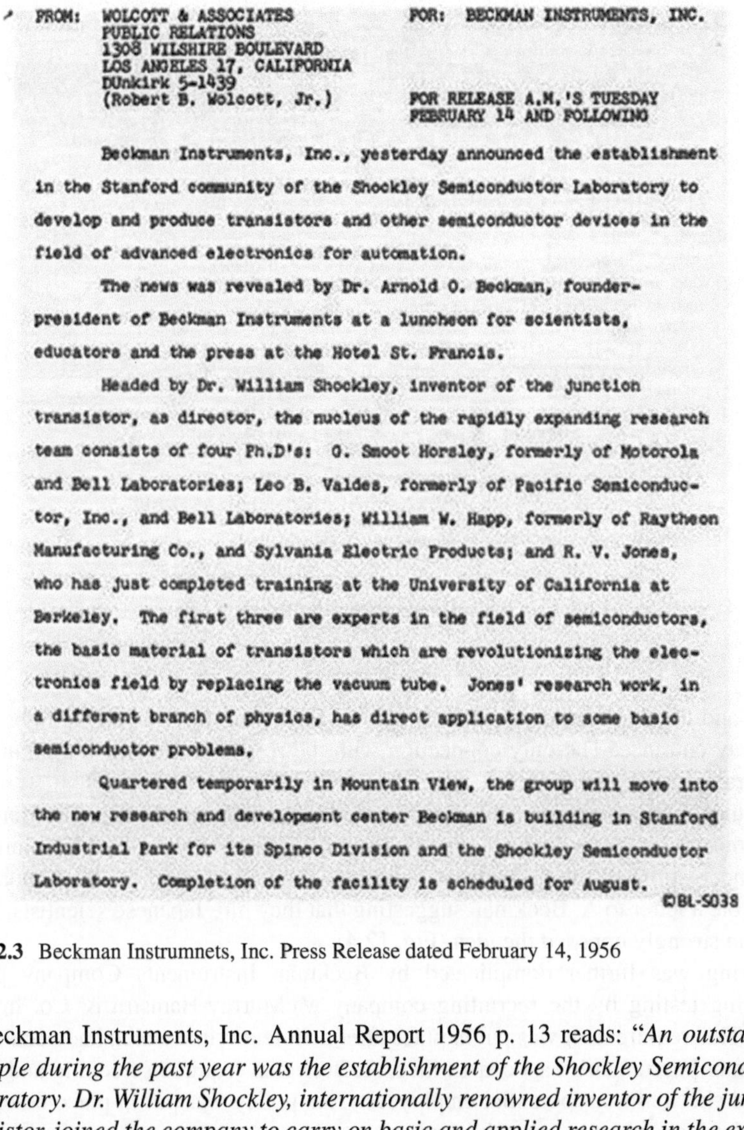

FROM: WOLCOTT & ASSOCIATES   FOR: BECKMAN INSTRUMENTS, INC.
        PUBLIC RELATIONS
        1308 WILSHIRE BOULEVARD
        LOS ANGELES 17, CALIFORNIA
        DUnkirk 5-1439
        (Robert B. Wolcott, Jr.)   FOR RELEASE A.M.'S TUESDAY
                                           FEBRUARY 14 AND FOLLOWING

        Beckman Instruments, Inc., yesterday announced the establishment
        in the Stanford community of the Shockley Semiconductor Laboratory to
        develop and produce transistors and other semiconductor devices in the
        field of advanced electronics for automation.

        The news was revealed by Dr. Arnold O. Beckman, founder-
        president of Beckman Instruments at a luncheon for scientists,
        educators and the press at the Hotel St. Francis.

        Headed by Dr. William Shockley, inventor of the junction
        transistor, as director, the nucleus of the rapidly expanding research
        team consists of four Ph.D's: O. Smoot Horsley, formerly of Motorola
        and Bell Laboratories; Leo B. Valdes, formerly of Pacific Semiconduc-
        tor, Inc., and Bell Laboratories; William W. Happ, formerly of Raytheon
        Manufacturing Co., and Sylvania Electric Products; and R. V. Jones,
        who has just completed training at the University of California at
        Berkeley. The first three are experts in the field of semiconductors,
        the basic material of transistors which are revolutionizing the elec-
        tronics field by replacing the vacuum tube. Jones' research work, in
        a different branch of physics, has direct application to some basic
        semiconductor problems.

        Quartered temporarily in Mountain View, the group will move into
        the new research and development center Beckman is building in Stanford
        Industrial Park for its Spinco Division and the Shockley Semiconductor
        Laboratory. Completion of the facility is scheduled for August.
                                                                    ©BL-S038

Fig. 12.3 Beckman Instrumnets, Inc. Press Release dated February 14, 1956

Beckman Instruments, Inc. Annual Report 1956 p. 13 reads: "*An outstanding example during the past year was the establishment of the Shockley Semiconductor Laboratory. Dr. William Shockley, internationally renowned inventor of the junction transistor, joined the company to carry on basic and applied research in the exciting new field of semiconductors.*"

Bell Laboratory scientists were willing to consider Fullerton but rejected Palo Alto. At that time RCA Laboratories and Bell Telephone Laboratories were very likely the best research organizations in the nation. They were well equipped, and surrounded by the best schools and universities. On the contrary, Santa Clara County and "*California near Stanford University*" was a rural farming community with no direct long-distance phone, while Mountain View had fewer than 6000 residents in

This morning I had occasion to speak to my old friend Dr. Harry Kelly, now with the National Science Foundation, in respect to another matter. I told Kelly of our activities and problems. He suggested recruiting scientists from Japan and offered to help in this. Kelly was the scientific representative of our occupation forces in Japan for several years after the war and is very highly regarded by both the Japanese and the Americans.

I have discussed this with Tom McCraney, who finds no legal barriers to recruiting Japanese and does not think that Immigration problems would be at all serious.

I am very enthusiastic about this possibility since the Japanese have been quite active in theoretical and experimental work in the field of semiconductors and transistors, although I think the facilities and leadership have not been comparable to those in this country. It seems to me that this could help us very materially in building up a technically strong group of willing workers.

As matters stand now I have asked Kelly to explore the possibilities here but not to mention us in this connection.

If it is agreeable to you, I shall remove this constraint and say that we have a real interest in this and ask Kelly to come up with some specific names that we can look into.

Fig. 12.4 Shockley's letter to A. Beckman suggesting that they hire Japanese scientists. (November 1955)

1955, and the San Antonio road was a gravel track with just one signal light at the highway entrance. The only connection with the rest of world was a train and the San Francisco airport.

Rather than advertising, Shockley started personally contacting his friends in academia and industry. After six months recruiting Shockley realized how much he had underestimated the problem of finding qualified personnel for his Laboratory. He wrote a letter to A. Beckman suggesting that they hire Japanese scientists, but F. Terman strongly opposed the idea (Fig. 12.4).

Hiring was further complicated by Beckman Instruments Company policy requiring testing by the recruiting company McMurray-Hamstra & Co. in New York. This requirement was put in place after a violent crime which had occurred in Beckman Instrument's east coast office, A. Beckman required screening and background checks for all potential employees. This policy was put in place before the Shockley Laboratory was established.

In a letter to Beckman dated February 20, 1956 Shockley wrote. "… *early next week you would have [the] opportunity to meet two candidates we are interviewing. One is Robert Noyce who has gotten a high recommendation from Schaeffer in the New York McMurry-Hamstra office. The other is Dean Knapic who should be head of our mechanical work …*"

On April 1, 1956 two new employees, Dean D. Knapic and C. Sheldon Roberts joined the Laboratory. Knapic was recommended to Shockley by his Western Electric friend Rudy Malina. Knapic brought with him two engineers, Julius Blank and Gene

Fig. 12.5 Dean Knapic, W.
Shockley, and Smooth
Horsley (1955)

Kleiner. Shockley viewed them as people who would do equipment assembly and production. Dean Knapic was a man who already knew everything. When Shockley asked *"Do you think you could do this?"* Knapic's response was always *"yes, of course"*, no matter what the job was.

The metallurgist Shockley was looking for was found in MIT's list of alumni. Sheldon Roberts had graduated in 1952 from MIT with a degree in metallurgy and was working at the Dow Chemical Company. His thesis advisor provided Shockley with an excellent recommendation.

Self-confident and overly ambitious, Robert Noyce was looking for a new opportunity after he had run into conflict with W.E. Bradley at Philco Corporation. His brother was an assistant professor at UC Berkeley, so he replied to the advertisement in Chemical and Engineering News. Noyce was also an MIT graduate; he defended his thesis *"A photoelectric investigation of surface states in insulators"* on September 1953. In the thesis he investigated the properties of magnesium oxide. He joined the Philco Corporation in Lansdale, and worked on equipment development for liquid indium sulfate jet stream etching of surface barrier transistors.[2] Noyce was interviewed on February 24, 1956 in Mountain View and immediately signed a job acceptance letter with a note that he would resign from Philco not later than April 9, 1956. When asked by Shockley why he was leaving Philco, Noyce stated *"the supervisor has not too much to say."* As time proved later, Shockley should also have paid attention to Noyce's comment, which Shockley recorded as *"intrigued by LBV (Valdes)."* Robert Noyce joined Shockley Laboratory on April 15, 1956.

On April 10, 1956 Shockley sent a letter to A. Beckman *"... none of these replies (to the advertisement) have been proven to be exciting ... we also need [to] add managerial talent along business lines to my organization and it would most probably*

---

[2]Contrary to lore, Noyce was not involved in device development; he was assigned to the project of optical measurement of the thickness of the etched base layer.

Fig. 12.6 Shockley Semiconductor Laboratory of Beckman Instruments, Inc., at 391 South San Antonio Road, Mountain View (1956)

come from Beckman Instruments. Maybe this suggests, as Mr. Shaw said, 'with your brains and my looks' ..." Beckman did not reply to this letter.

Shockley had no time to work on the engineering problems; all his time was taken up by recruiting and traveling to meet or interview potential employees. Shockley's idea of recruiting people who could make their living as research scientists but had the willingness *"to bring stuff to utility"* was unsuccessful. After more than 12 months of Shockley's employment with Beckman Instruments, Shockley Laboratory ended up with 32 employees (Fig. 12.6).

Apart from D. Knapic, who was 35 years old, none of the engineers and scientists were older than 30. To set up the right R & D group, there had to be a reasonable balance between senior and junior personnel. There is an energy and enthusiasm that young people can bring to a research group, but there is also a calmness and wisdom that the more senior person can bring. A senior member of a research group plays an indispensable role in development teams. Many fresh graduates, as with many young people, already know everything better than anyone else. There was no one in the Shockley Semiconductor Laboratory who could calm the unrealistic expectations of young and impatient employees. Apart from Valdes and Noyce, who were exposed to transistor physics, all the other members of Shockley's team, to use R. Jones terminology, *"knew bupkis about semiconductors."*

I found an interesting note in Shockley's papers dated May 31, 1956. The note reads: *"Hoerni was the only one who asked what type of work he would be doing and did not ask questions about salary."*

There was one employee who played a pivotal role in the fate of the Shockley Semiconductor Laboratory—Gordon E. Moore. Moore received his Ph.D. degree

EMPLOYEES
8/28/56

1. W. Shockley		22. W. Pleibel
2. L. Valdes		23. E. Kleiner
3. W. Happ		24. V. Grinich
4. S. Horsley		25. G. Troyer
5. V. Jones		26. S. Lee
6. R. Grunewald		27. M. Asemissen
7. T. Zinn		28. C. Sah
8. L. Bolender		29. J. Hoerni
9. A. Pretzer		30. K. Jacobsen
10. S. Roberts		31. Bill Stansbeary
11. C. Himsworth		32. D. Farwell
12. H. Breen		
13. D. Knapic		
14. R. Noyce		
15. R. Wagner		
16. J. Blank		
17. G. Moore		
18. D. Allison		
19. G. Stout		
20. J. Last		
21. W. Gadbury		

Fig. 12.7 List of employees at the Shockley Semiconductor Laboratory as of August 28, 1956. (The left column is the employee number in order of hiring date)

in chemistry and physics from Caltech in 1954. The title of his thesis was "*Infrared Studies of Nitrous Acid, the Chloramines and Nitrogen Oxide.*" He spent the next 16 months as a postdoc at the Applied Physics Laboratory of Johns Hopkins University. Moore wanted to return to California where he had grown up in Pescadero, near San Francisco.

On February 2, 1956 Horsley sent a letter to Shockley who was in New York: "*Vic Jones this afternoon went over to Livermore and obtained enclosed list of names from their personal files. All of the individuals on the list were invited to visit Livermore for prospective employment.*" The list contained the names of seven chemists, one whom was Gordon E. Moore. Moore was also interviewed by Lockheed during February 16–23, 1956 and Dr. Spencer passed Moore's name to Shockley. Based on a few phone conversations with Shockley, Moore was immediately offered a job (Fig. 12.8). Moore joined the Shockley Semiconductor Laboratory on Monday, April

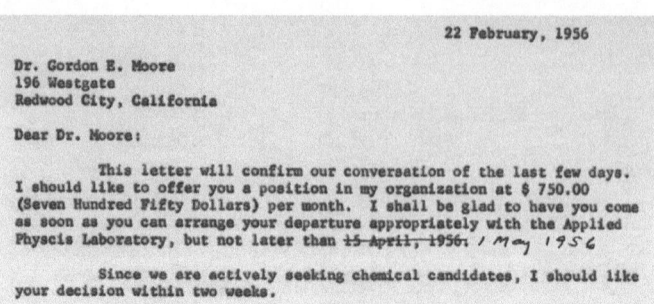

Fig. 12.8 Job offer to G. E. Moore

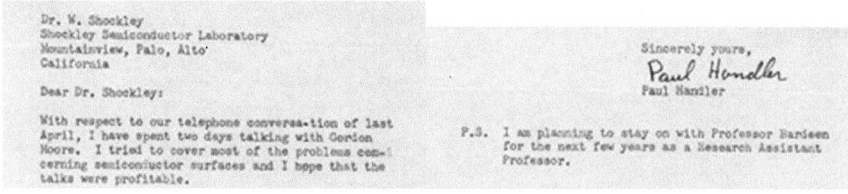

Fig. 12.9 G. Moore's training visit in J. Bardeen's group at the University of Illinois (May 1956)

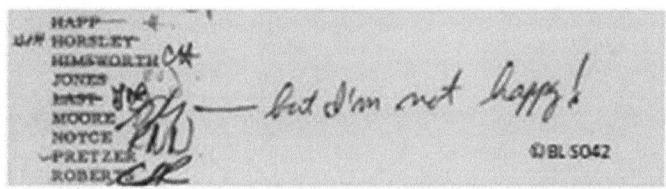

Fig. 12.10 A typical comment by G. Moore on Shockley Laboratory's interoffice memos

16, 1956. Shockley then dispatched Moore to the University of Illinois to learn about the semiconductor surface states (Fig. 12.9).

For some reason, or rather for many reasons, from the beginning of his employment with Shockley Laboratory, Gordon Moore had a different attitude than other employees. By the beginning of January 1957 he was unhappy with almost everything. Internal company memoranda contain Moore's comments, such as "*had some reservation and suggestion*" or "*will attend but not happy about schedule*" or something like Fig. 12.10.

On March 13, 1956 Shockley placed a new ad in the New York Herald Tribune and The New York Times:

Shockley Semiconductor Laboratory of Beckman Instruments, Inc. has opening for qualified electronic engineer with research and development experience in semiconductor device design and/or application, computer circuitry or related fields. Some of many benefits of joining our firm include (1) the opportunity to grow with present small organization (2) located in California near Stanford University (3) where you will work in close contact with scientists of varied disciplines. For interview call MURRAYHILL 2-7430.

Yet another, typical Shockley, advertisement[3] was placed in Proceedings of the IRE.

AKABSQIMEBR AMFEMAAQR:

E YL KIIJEMF DIQ RATAQYK EMFAMEIOR NAINKA SI CAREFM
KYZIQYSIQU YMC SARS APOENLAMS YMC SI GAKN EMTAMS MAV
RALEBIMCOBSIQ CATEBAR DIQ BILNOSAQR.

V = RGIBJKAU
RGIBJKAU RALEBIMCOBSIQ KYZIQYSIQU
391 RI. RYM YMSIMEI QC.
LIOMSYEM TEAV*
BYKEDIQMEY

* MAYQ RSYMDIQC

(For appointments during New York meeting call Murray Hill 2-7430)
To decipher: Replace Y by A, E by I, I by O, etc., and Z by E, B by C, C by D, etc.

©BL-S202

Note the phone number, Shockley Laboratory did not have long-distance phone access. Calls were redirected to Bell Laboratories where Betty Sparks kept contacts for all applicants and forwarded information to Ms. C. Himsworth at Shockley Laboratory. A list of almost 300 names who responded to this ad were, after initial contacts, labeled with status "*Dead*" or "*Limbo*." About 90 names had a status "*We have the ball*" or "*He has the ball*." The effort yielded only five new employees. Shockley realized that recruiting qualified people to the "southern edge of Palo Alto" was the main problem he was facing. After this recruiting fiasco, anyone with a reasonable qualification was hired instantly.

Despite the difficulties in finding new employees, Shockley's behavior contrasted with the aura created about him. On January 17, 1957 R. V. Jones told Shockley that he would like to return to academia and resign from Shockley Laboratory. Shockley wrote a letter to Roman Smoluchowski with a warm recommendation and appreciation of Jones' work at Shockley Laboratory. On Wednesday, February 21, 1957 Shockley organized a farewell lunch for Jones. Jay Last, the youngest member of the R & D group, was also not clear about his professional determination and asked Shockley to help him to find work in Germany. On April 26, 1957 Shockley sent a letter to Beckman and asked him to arrange Jay Last's employment in the Beckman Instrument, Inc. Munich office (Fig. 12.11).

Recruiting, however, was not Shockley's only problem. At that time there was no source of commercially available silicon or germanium. Shockley assumed that he would be able to purchase manufacturing equipment from Texas Instruments (Fig. 12.12).

Shockley wanted to make a "*research effort in the field of transistor and other semiconductors,*" but he had no semiconductor material. Instead of focusing on the

---

[3]Jack Ryder sent a reply in the same code as mentioned in the footnote: "*Having been in this so long you should know that electronics is a noun and not an adjective.*"

One of the young men in my organization, a physicist by
background with experience in optics, and more recently in semiconductor physics, would like to spend a year in Europe before
settling down more permanently in one place. Munich appeals to
him. I think he is as much interested in travel and cultural aspects of the work as in its actual nature.

I am in sympathy with his wishes and wonder if there is
a possibility of working out a transfer for a year so that his con-
tinuity in the Beckman organization would be preserved.

©BL-S189

Fig. 12.11 Shockley's letter to Beckman (April 26, 1957)

5. Crystal Grower from Texas Instruments

I have been in communication with Pat Haggerty to find
out whether or not Texas Instruments would sell us a crystal
growing machine. Haggerty says that Texas has been approached
a number of times and never sold a machine. However, he has
expressed willingness to query his board of directors and see
whether or not they might agree in this case. Their machines are
suitable for growing crystals either of silicon or germanium.
Haggerty does not know what the price would be but made a rough
estimate that it might be on the order of $30,000 without the
associated high frequency heating equipment. It is my belief that
we could build one for considerably less with the aid of Charles
Bittman, if we take him on.

Fig. 12.12 Memo to A. Beckman (September 1955)

development of devices, he needed to change strategy and develop and construct
manufacturing equipment.

In order to do some preliminary work on the four-layer diode and field effect
device, Shockley asked Jack Morton to provide silicon material from Bell Laboratories. The first batch of material was sent to Mountain View in April 1956 and
continued until the year's end.

On October 12, 1955 Pat Haggerty notified Shockley that Texas Instruments would
not sell the "crystal puller" (Fig. 12.13).

Shockley's friend, Jack Morton, was a complicated personality. He was a driving
force behind the Type-A transistor. J. Goldey described Jack as follows: *"He was
hard working, hard playing, hard drinking. He was a driven man, no question about
it. He was a man of great vision. He understood the impact of the diffusion and
enjoyed the things he did too. I guess everybody did, but Jack drove it. He was very
good at getting resources. He could be a very tough guy too. I think he may be
underappreciated because of some of his bad calls, but I think he was very important
to that whole effort. He and Shockley got along very well together"*(sic).

In mid-1956 the employee assignment was as follows:

TEXAS INSTRUMENTS

October 12, 1955

Dr. William Shockley
Beckman Instruments, Inc.
North Fullerton Road
Fullerton, California

Dear Bill:

It was nice to talk to you last week. I am sorry to
have taken so long to reply to your request on crystal pullers, but,
frankly, we were doing some wrestling with ourselves to see
whether we felt we could logically sell pullers or not. Unfortunately,
it is with regret that I must advise you that we still feel it would be
unwise for us to sell the pullers. Should we ever change our minds
on this score, I certainly should advise you as promptly as possible,
by which time, of course, you may not be in the least interested.

I certainly wish you the best in this new endeavor of
yours. Don't hesitate to stop in to visit us in Dallas whenever you
come this way.

Sincerely,

P. E. Haggerty

PEH/nj   ©BL-S212

Fig. 12.13 P. Haggerty's letter to W. Shockley (October 12, 1955)

Physics Group: Shockley, Horsley, Noyce, Jones, Happ, Last
EE group: Valdes, Grinich, Sah
Chemistry: Moore, Blank, Madden
Metallurgy: Roberts
Theoretical Physics: Hoerni
Manufacturing: Knapic, Kleiner, Blank

Leo Valdes, who was in charge of the crystal growing equipment, was spending all his time planning to build crystal growing equipment. He and Shockley set up the initial concept of the equipment and Valdes was already ordering parts. He was the most experienced person. Valdes and Noyce did not get along because Valdes thought he knew more about semiconductors than Noyce. Valdes and Happ were the only people with some autonomy until Noyce arrived. Because Shockley put his money on Noyce, Valdes resigned from Shockley Laboratory in September 1956.

After Valdes left, the development of crystal growing equipment was assigned to R.V. Jones. After about six months' work on crystal growing equipment, Jones resigned. It seems that industrial research and development was not a subject the physicist who studied nuclear magnetic resonance at UC Berkeley would enjoy. When his ideas about zone refining were disapproved by Hoerni and Grinich, Jones

Fig. 12.14 Shockley & Jones's crystal growing apparatus (August 2, 1956)

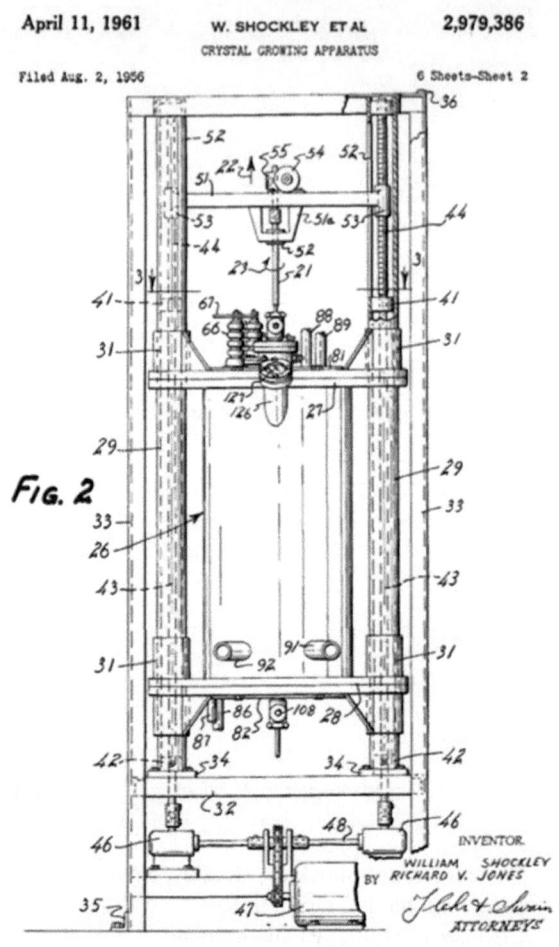

After Valdes and Jones, the third person working on the crystal growing equipment was C. Sheldon Roberts. Shockley sent him first to Bell Laboratories Metallurgical Group to learn about the metallographic properties of silicon crystal surfaces. Roberts did not like certain features of the equipment that had been designed by his predecessors. Shockley wanted to have the equipment up and running, while Roberts was proposing and making changes in the design. They argued almost constantly. In spite

complained about *"mental stagnation"* to G. Moore the next day and decided to separate from Shockley Laboratories. His work resulted in a patent application "Crystal growing apparatus" co-authored with Shockley and submitted to the Patent Office on August 2, 1956 (Fig. 12.14). Jones wanted to work on his primary interest, electromagnetic theory. In his opinion, Jones later stated: *"Shockley was a genius that was shortsighted because he had no use for magnetic resonance."* [4]

---

[4]R. V. Jones, interview by D. C. Brock, 2006.

Stanford Physicist Wins Share in Nobel Prize

William Shockley of Los Altos and two other American scientists were awarded jointly the Nobel physics prize today for their invention and development of the point contact transistor, which is making revolutionary changes in the radio and telephone industries.

The invention also has been a boon to hearing aids.

The physicists sharing the $38,633 in prize money with Shockley are Walter Brattain of Murray Hill, N. J., and Prof. John Bardeen of the University of Illinois.

On Stanford Faculty

Shockley, 46, is director of the Shockley semiconductor laboratory division of Beckman Instruments at Mountain View... engineering at Stanford.

A quiet, soft-spoken man, he lives with his second wife, Emmy, whom he married last Nov. 23, at 23466 Corte Via, Los Altos.

Wearing an Hawaiian sport shirt and slacks, he sat in the study of his home drinking cof-

the junction transistor has greatly increased the versatility and reliability of many electronic and communications devices.

Shockley was born in London, England, of American parents and came to the United States as a youngster. He attended elementary school in Palo Alto and high school in Los Angeles. In 1932 he received his bachelor's degree from California Tech and his doctor's degree four years later from Massachusetts Tech.

He joined Bell in 1936. During World War II he was director of research for the Navy's anti-submarine warfare operations research group and a consultant in the office of the Secretary of War.

He predicted the wider use of ... which, he said, will ... more cheaply as sales increase.

Shockley said there are "many new aspects to be developed." Present work might lead to "new kinds of devices in the field of science," he said, but he declined to elaborate.

Fig. 12.15 The San Francisco Chronicle, November 1, 1956 with the official UPI photo of William and Emmy Shockley

of these peripeteia, the equipment was working correctly by the end of January 1957, and produced the first silicon ingots with a diameter of approximately 8 mm.

One employee, Swiss born Jean Hoerni, did not work in "Building 1"[5] at San Antonio Rd. Shockley wanted to have his own mathematician, similar to "Van."[6]. Hoerni received his first Ph.D. degree in mathematics and physics from the University of Geneva, and the other from the University of Cambridge. Hoerni moved to Caltech in 1952 as a postdoc with Linus Pauling where he worked on the application of mathematical problems in physics and chemistry. Per Pauling's recommendation, he joined Shockley Laboratory in June 1956. His first assignment was to calculate Gaussian and error function concentration profiles. Because he did not want Hoerni to be distracted by activities in "Building 1," Shockley rented an apartment where Hoerni worked in isolation during the day. Jay Last used this place at night before he found his own place.

A major disruption for Shockley occurred on November 1, 1956. Early that morning when The San Francisco Chronicle was already out, he received a phone call that he had been awarded the Nobel Prize together with Bardeen and Brattain (Fig. 12.15).

For the next few days no one in Shockley Laboratory thought about business, and everyone forgot about the difficult, overbearing, and suspicious William Shockley.

---

[5]Beckman/Spinko on California Avenue in Palo Alto was "Building 2".

[6]Van was a nickname for Bell Laboratories' mathematician Rudy van Roosbroeck.

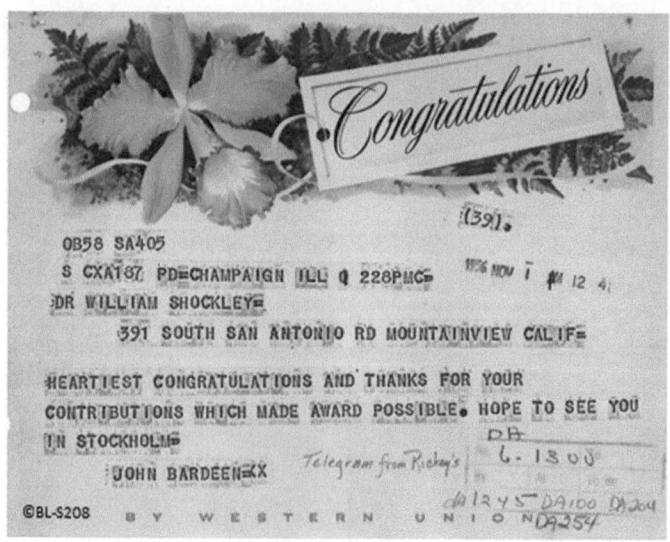

Fig. 12.16 J. Bardeen's congratulations to W. Shockley (November 1, 1956)

John Bardeen sent a telegram to Mountain View and Walter Brattain added in his telegram: "*Trusting you still sober and coherent at this stage of the game congratulation. Brattain, BellTelLabs*" (Fig. 12.16).

All Shockley Laboratory employees gathered in Ricky's restaurant for a lunch and congratulated their boss. That evening Shockley, Emmy, and May had dinner, and May added her recollection of bringing up Billy. She was very proud of her son! (Fig. 12.17)

On November 20, 1956 Shockley replied to Kelly's congratulations: "*Dear Mervin, it is hard for me to see as a research director and vice president in your position could have proceeded more effectively to get transistor out of a solid-state physics. The background of experience I had in vacuum tube area and talks you once gave me on the importance of electronic switching stimulated me to be alert to such possibilities. This was then followed by the freedom to work on subjects of my own choosing in the solid-state physics area. I hope that we shall have an opportunity to pat each other on the back over a drink before too long.*" (sic)

To his friend, Fred Seitz, Shockley replied on November 26, 1956 "*It was nice to receive your telegram and I would like to thank you for it. I felt that the award recognized the transistor work not only as a separate isolated entity but as a fact of large and prospering field of solid-state physics.*"

Shockley certainly deserved the Nobel Prize. The problem, however, was that the timing was unfortunate. Shockley Laboratory was understaffed, and it was running into financial and interpersonal problems which required immediate attention and action, but the leader was distracted. The Nobel Prize often has adverse effects on laureates and their future professional careers. In the first place Nobelists will find

Fig. 12.17 Congratulations to Shockley Nobel Prize from Shockley Laboratory employees

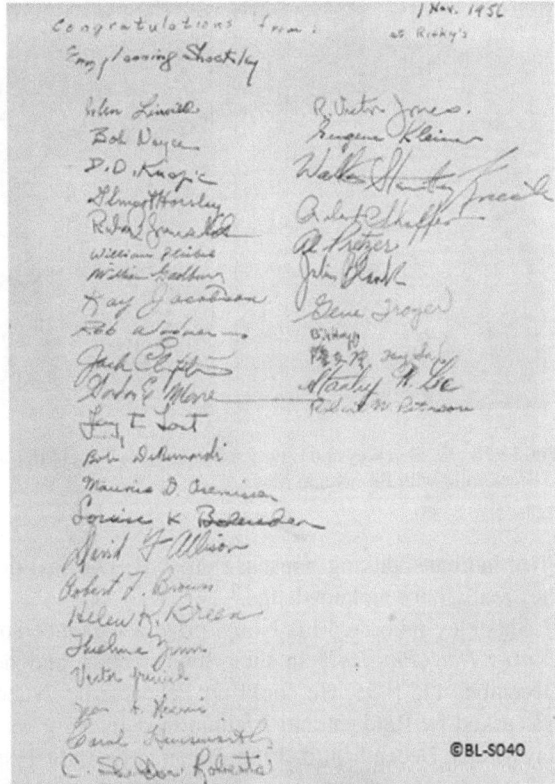

themselves on all kinds of committees, and invited to make presentations and attend "show" gatherings.

Shockley, whose work stood out for many years as extraordinary, confirmed his credentials for the Nobel Prize that he shared with Bardeen and Brattain. The fact is that Shockley's theory of the p-n junction and junction transistor, which dominated the market at that time, and the field effect transistor which dominated the market later, alone deserved the prize. But when you are famous it is hard to work on small problems. One such problem was reduced productivity.

Perhaps Alan Chynoweth provided the best explanation: *"The day the prize was announced in Bell Laboratories Arnold Auditorium, Brattain, practically with tears in his eyes, said, I know about this Nobel-Prize effect and I am not going to let it affect me; I am going to remain good old Walter Brattain. Well I said to myself, that is nice. But in a few weeks, I saw it was affecting him. Now he could only work on great problems."*

If someone wins the Nobel Prize, he is immediately pounced upon by all kinds of people who ask him all kinds of questions on all kinds of subjects which he knows nothing about. It is very difficult to resist the temptation to reply—to give an answer on subjects which one knows nothing about. The good reader will have seen Nobel

Fig. 12.18 W. Shockley and Emmy rehearsing his Nobel prize acceptance speech, and on the flight to Stockholm with Emmy and May

Prize laureates talking nonsense on some political question or other, about which they really have no knowledge.[7]

Shockley rehearsed his Nobel prize acceptance speech *"Transistor Technology Evokes New Physics"*[8] in their living room, and delivered it in Stockholm on December 11, 1956. He could not rehearse his Nobel Banquet speech, which he was asked by Bardeen and Brattain to deliver for recipients of the Nobel Prize in Physics. on December 10, 1956. He quickly studied past volumes of *Les Prix Nobel* and stated: *"My difficulty was not in finding the correct sequences but in finding them so often so well phrased. My experience reminded me of T.S. Elliot's frustration in the second of his "Four Quartets". Elliot writes: And what there is to conquer by strength and submission, has already been discovered once or twice, or several times by men whom one cannot hope to emulate ... gratitude is expressed both for the award of the prize and for the contributions of earlier contemporary scientists and colleagues."*

Shockley, Emmy, and May arrived in Stockholm on Sunday and had a day to rest in a Grand Hotel suite (Fig. 12.18). The Nobel festivities started on Monday, December 10, 1956 at 4:30 PM. All the Nobel laureates entered the Concert Hall with trumpet fanfares, and Prof. Erik Rundberg read the charge starting with *"The summit of Everest was reached by a small party of ardent climbers."* Then Shockley went first, followed by Bardeen and Brattain to shake hands with King Gustav VI Adolph and receive the gold medals. Emmy and May Shockley with Jane Bardeen and Karen and Bill Brattain watched the ceremony from the front row.

---

[7]I. Giaever, *"I am the smartest man I know"*, World Scientific, 2017.

[8]https://www.nobelprize.org/prizes/physics/1956 did not include Shockley's speech in the web site (accessed on 10/01/2020).

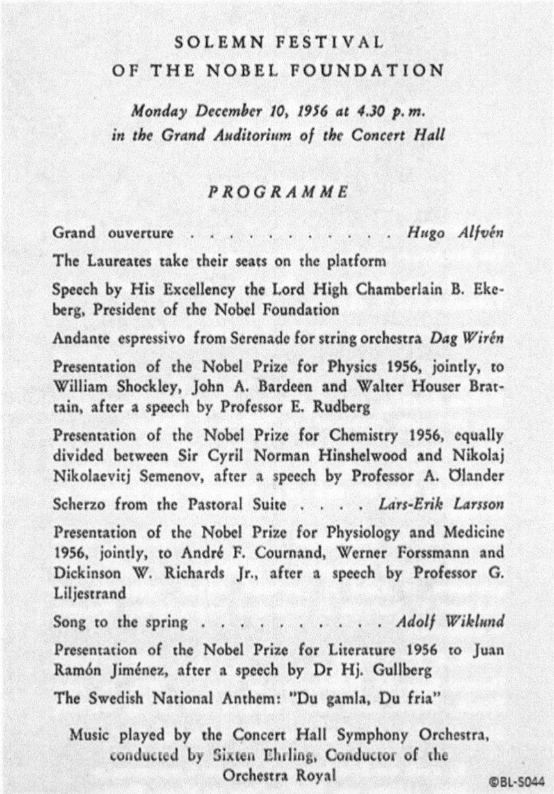

Fig. 12.19 Program of the
1956 Nobel Prize ceremony

SOLEMN FESTIVAL
OF THE NOBEL FOUNDATION

Monday December 10, 1956 at 4.30 p.m.
in the Grand Auditorium of the Concert Hall

PROGRAMME

Grand ouverture . . . . *Hugo Alfvén*

The Laureates take their seats on the platform

Speech by His Excellency the Lord High Chamberlain B. Eke-
berg, President of the Nobel Foundation

Andante espressivo from Serenade for string orchestra *Dag Wirén*

Presentation of the Nobel Prize for Physics 1956, jointly, to
William Shockley, John A. Bardeen and Walter Houser Brattain, after a speech by Professor E. Rudberg

Presentation of the Nobel Prize for Chemistry 1956, equally
divided between Sir Cyril Norman Hinshelwood and Nikolaj
Nikolaevitj Semenov, after a speech by Professor A. Ölander

Scherzo from the Pastoral Suite . . . . *Lars-Erik Larsson*

Presentation of the Nobel Prize for Physiology and Medicine
1956, jointly, to André F. Cournand, Werner Forssmann and
Dickinson W. Richards Jr., after a speech by Professor G.
Liljestrand

Song to the spring . . . . *Adolf Wiklund*

Presentation of the Nobel Prize for Literature 1956 to Juan
Ramón Jiménez, after a speech by Dr Hj. Gullberg

The Swedish National Anthem: "Du gamla, Du fria"

Music played by the Concert Hall Symphony Orchestra,
conducted by Sixten Ehrling, Conductor of the
Orchestra Royal

©BL-S044

When the Swedish national anthem closed the ceremony, the new Nobelists returned to the Grand Hotel. Bardeen and Brattain with their guests ordered champagne in the hotel dining room while Shockley and Emmy roamed in the hotel lobby alone. After midnight Brattain tried to cheer them up and invited them to the party table. For a short time, they all enjoyed the glory of their distinction.

On way back from Stockholm, Shockley visited the German office of Beckman Instruments, Inc. Later, after Shockley's visit, followed by dinner at a Bavarian restaurant in Munich, Nils Liljenström, the Swedish representative of Beckman Instruments, Inc., stated that: *"Shockley was now entitled to be a full conductor and no longer a semi–conductor"* (Fig. 12.20).

When the festivities had calmed down and Shockley finally returned to work, he found the business in disarray. The Shockley Laboratory planned expenses for the first year to be around $500,000, but by the end of 1956 they actually exceeded $1 million. Spinco building construction and total research and development expenditures exceeded the limit which A. Beckman had set up. In addition, the company suffered losses from mishandled government contracts. By the beginning of 1957,

Fig. 12.20 *"Shockley was now entitled to be a full conductor and no longer a semi–conductor"*

Beckman Instruments, Inc., stock had lost almost 6% of its total value. Unscheduled development and construction with frequent changes to the crystal growing equipment, not to mention a failed attempt to use a commercially available diffusion furnace (Burrell furnace model 2957BT), significantly increased expenditure and caused time delays.

To invent a new device and provide a proof of feasibility is very exciting and creative. To set up the manufacturing environment to produce the same device day after day is tedious, much less creative, and often uninteresting work. Shockley Laboratory was developing not only a new, complicated device; they also needed to develop the manufacturing equipment. In Bell Laboratories, Shockley worked with highly educated scientists and specialists, probably the top people in the nation. He could just ask an appropriate department to develop what was needed. In California all needed to be done from scratch with employees who were novices in the field. Product development involved not only the physics, but the entire process of researching, designing, creating, marketing, and selling new products. Research and development are essentially the first steps in developing a new product, but product development is not exclusively research. To set up the manufacturing of any semiconductor device is tedious and often rather dull work, interfacing with people with very different qualifications. Shockley, was completely unprepared for these challenges.

The silicon crystal pulling machines and diffusion furnaces used in the research environment in Bell Laboratories were rather proof of feasibility than real manufacturing equipment. The tools used by germanium transistor manufacturers were inadequate for silicon devices. Shockley Laboratory was developing and assembling their own furnace for zone melting, diffusion, and crystal growing equipment.

The commercially available Burrel furnace was just a quartz tube with a wound resistive heater which evaporated at high temperature. The Burrel equipment had to be completely re-built with an expensive Swedish-built Kanthal heating element REH 4-60, Fisher & Porter Co. Tri-Flat (low-flow rate) flowmeters, and a Wheelco 407 controller. The cost of the tool tripled.

Arnold Beckman instituted cost-cutting measures. Shockley used his own money to pay the Patent Office fee for his and Jones' patent application for the crystal grower (U.S. Patent 2,979,386) and two additional patents. To find new funding for the Laboratory, Shockley turned to his old friends. The military funded four projects over the years 1957–1960: AF19(604)-5524 Cambridge Research Center, NONR2934(00) Naval Research, AF33(616) Wright-Patterson, and DA36-039 SC 85239 Signal Corps. All of them targeted the four-layer diode because of the enormous interest in this device. Motorola and Philco had already ordered the first available devices for "detonator firing applications", while AVCO wanted to use it in a ring counter, and MIT was developing core memory switching elements based on the four-layer diode. From now on Shockley and his laboratory had no choice, they needed to develop and deliver the four-layer diode, which was very complicated and difficult to manufacture with the technology available at that time.

Jean Hoerni was increasingly frustrated to work alone. Hoerni had the same kind of abrasive personality as Shockley; he was very sharp and was the brightest of all the scientists and engineers working for Shockley at that time. Hoerni wanted to be an experimentalist and directly involved in the diffusion experiments. In October 1956 he released a Technical Memorandum titled "Solution of the diffusion problem" where he calculated the Gaussian and error function concentration profiles and a second memo titled "Diffusion of concentration profiles for Al and P." He was literally begging Shockley to move him to Building 1 and include him in the group's work.

Sheldon Roberts, who wanted to design his own micromanipulator and crystal grower, got into an argument with Shockley. Shockley wanted to use parts from a commercially available metric micrometer, while Roberts wanted to design a more sophisticated instrument.

Both Hoerni's and Roberts' frustration reached the tipping point. On Thursday, February 28, 1956 they communicated to S. Horsley their intention to resign. Horsley immediately talked to Shockley who on Friday March 1, 1957 talked to R. Noyce, and then to Hoerni and Roberts. Shockley wrote this note: "*JH would not continue as theoretician.*"

On April 3, 1957 Shockley re-assigned the group as follows:

NPNP Switching Diode Program (Horsley, Allison, Roder)

Assembly: Brown

Diffusion, Evaluation and Control: Hoerni

Wafer Measurement, data interpretation: Last

Soldering, dicing: Roberts

Packaging and Fixtures: Knapic

Capacity effects: Sah

Field Effect Transistor (W. Shockley)

Fabrication and evaluation: Farwell, Parker, Pleibel, Tavares

Circuit application: Grinich, Paterson

Diffusion: Noyce, Blank, De Bernardi

Masking, etch, wax: Sello, Brown

Gallium diffusion, slicing: Fok, Lewis, Smith, Wagner

Crystal Growth: Moore, Ford, Grunwald, Clifton, Madden

Exploration: Noyce, De Bernardi

Any engineer who worked in a semiconductor Fab should now ask the question: How was it possible for a group of novices hired from February to June 1956, working in a facility which did not even have enough electrical outlets eight months earlier, was able to do work on diffusion, etch, and masking when they also needed to develop and assemble the manufacturing equipment, and when they did not have their own silicon substrates.

The answer to this question is quite simple. Shockley Laboratory was copying work and experiments performed in Bell Laboratories. Shockley was in almost daily contact with people working at Bell Laboratories. He was getting the diffusion-based

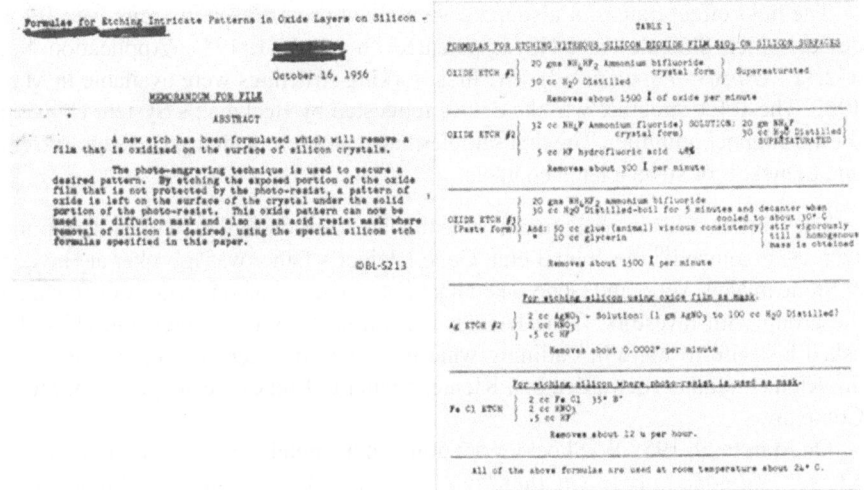

Fig. 12.21   One of the Bell Laboratories "cookbook" recipes

processes from Bell almost like a cookbook. John Moll[9] described the situation eloquently like this: *"Jim M. Goldey was kind of worried about that, that if we are giving out everything that we've struggled to get in almost cookbook form; which meant that it would save an awful lot of time on how you do it. Even the cleans and etches and things like that. He knew exactly how much of this or that. And if it was a different temperature than room temperature, what temperature. We checked with the management at Bell, and they said, "Yes. Go ahead and help him." And I think there was a certain amount of—well, Shockley left in a friendly way, and so there was an attempt to help him."*

Bell Laboratories' memos detailed the diffusion, etch, and masking that were made available to Shockley Laboratory engineers. Hoerni, Last, and Roberts used Bell Laboratories' data to their advantage, and in two months they delivered excellent work. Hoerni interpreted phosphorus (1100 °C/30 min) and boron (1200°C/60-120 min) diffusion data for the NPNP project and they converted the closed tube diffusion to the open tube, as suggested by Frosh at Bell Laboratories (Fig. 12.21).

Hoerni and Last published an internal report on April 15, 1957 stating: *"Use O_2 instead of wet N_2 as a carrier gas during the boron diffusion. According to the BTL reports of Frosh et al. this leads to more successful oxide masking"* and *"switch from B_2O_3 to BCl_3 as a boron source."* Hoerni and Last continued: *"The principal point we need to discuss is: are we making optimum progress towards production or would we get to production more rapidly by starting on a program of making Chinese copy of present method used at BTL."*

---

[9]J. Moll in an interview with A. Goldstein, IEEE 1993.

The field-effect transistor also made significant progress, on the improved JFET device which Shockley and Noyce patented on April 11, 1957 (Application No. 652,11 *"Transistor structure."*). The first working structures were available in May 1957. The JFET structure was the device requested by Beckman's System Division for the chopper amplifier. The first samples had a problem with high leakage current due to nickel- or silver-plated contacts.

The April 3, 1957 re-assignment came too late. Gordon Moore, with underground secrecy, organized[10] the coup d'état. Gene Kleiner's father was a broker at Hayden & Stone in New York, and Gene asked his father to find someone who could connect the group with investors. Gene Kleiner was afraid to write a letter himself so he asked his wife to write it, outlining what the group of future defectors wanted to do. Kleiner's father forwarded Ms. Kleiner's letter to Bud Coyle at Hayden & Stone Company.

On March 29, 1957 W. Shockley set down in his notebook: *"... surprised to find Kleiner had gone to east. On Tuesday 26 gone to LA for metal show. I thought he will be gone 3 days. Actually, he will be gone altogether 9 working days. Production chart is 4 weeks out of date ..."* In fact, Gene Kleiner went to New York to meet with Coyle and provide more information for Hayden & Stone's analyst.

As the outcome of the Hayden & Stone decision was not yet known, Moore opened a second front to get rid of Shockley. Gordon Moore called A. Beckman and filled him in with his view of Shockley Laboratory's problems. Beckman agreed to meet with people identified by Moore in San Francisco at the end of May 1957. Not all of the future defectors were present; Robert Noyce was not part of the group.

Moore, the self-appointed group leader who had erred into thinking rather highly of himself, complained that Shockley had put Horsley in charge of the NPNP diode project, that Noyce and De Bernardi were working on a device which no one knew about,... and that Shockley was oppressive. At the end of the meeting, Moore presented Beckman with the only acceptable solution: Shockley must be removed from his post but might work as a consultant. The group ultimatum was summarized as *"unless something is done, they would all resign."* A. Beckman did what no good businessman should do, he made a promise without knowing where both sides stood.

On May 23, 1957 Beckman invited Shockley and his wife for a dinner at the Jack Tarr Hotel in San Francisco. Beckman did not disclose to Shockley his meeting with the defectors but shared with him his worry about the situation in the Laboratory and acknowledged Moore's phone call and the threat of the resignation of personnel unsatisfied with the Shockley management style.

Shockley was in shock and asked Beckman why G. Moore had not talked about his concern with him in the first place. Shockley just could not understand that his management approach *"tell me what you do and I will tell you how to do it better"* was not easily digestible by fresh Ph.D.'s who thought they knew better. Shockley's contract with Beckman would run out on September 7, 1957 and he stated clearly that he would leave if he could not remain in the director's position.

---

[10] After split with G. Moore, J. Hoerni used the expression *"the mastermind."*

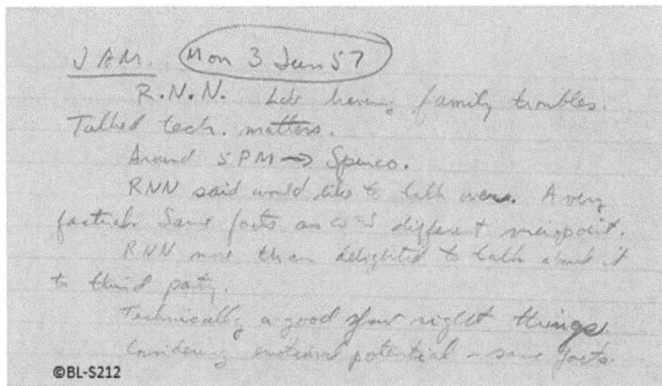

©BL-S212

Fig. 12.22 Notes in Shockley's notebook (June 3, 1957)

He argued with Beckman that all of the rebels had accepted job offers, and that they had all been looking for a new opportunity because they were unsatisfied with their previous employment. They had been offered work with state-of-the-art technology, and Shockley had encouraged them to grow in their careers, seek patent applications, and publish their work. Jones and Noyce had already submitted their patent, and Jay Last had just submitted his first paper with Shockley in the Physical Review [11]. Shockley was puzzled by Beckman's rounded statements; he thought he had a good relationship with him. Shockley clearly did not understand where the real problem lay.

Beckman concluded that Noyce was not among those who would resign and suggested to Shockley to work the problem out with him. Shockley trusted Noyce but told Beckman: *"Noyce is good technically but Horsley is [the] better physicist."*

Shockley talked with Noyce and when he asked him for help, he wrote down on Friday, May 31, 1957 in his notebook: *"Noyce more than delighted to talk to third party"* (Fig. 12.22).

Robert Noyce met with the dissidents on the evening of Monday, June 3, 1957. On June 5, 1957 Noyce presented Shockley with the changes suggested by the "Group":

- Noyce could work as a team with Shockley
- if they went ahead, 3 people would leave
- there was a major objection against W. W. Happ
- the company should sell the silicon crystal
- an experienced business manager must be hired
- decisions would be made by an interim committee including the new manager, Noyce, Knapic, and Moore (if he stayed with the company).

There was no discussion about the scope of the different projects. Noyce also informed Shockley that the reactions of individuals were unstable and that the

---

[11] W. Shockley, J.T. Last *"Statistics of the charge distribution for a localized flaw in a semiconductor"*, Phys. Rev. Vol. 107 (1957), pp. 392-396. Submitted on April 5, 1957.

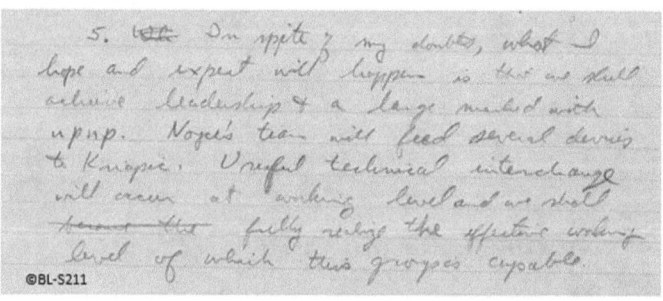

Fig. 12.23 W. Shockley notebook dated June 5, 1957 after the meeting with R. Noyce (*it took a lot of doing to go behind my back*)

Fig. 12.24 The expected *"useful technical interchange"* in Shockley Laboratory never occurred

"Group" was doing work on a copy of Bell Laboratories' mesa diffused transistor (Fig. 12.23).

Shockley responded to Noyce *"I believe it can be still made to go successfully. I would prefer [to] try to do so."*

On June 6, 1957, A. Beckman called Shockley. He asked about the meeting with Noyce and presented the compromised plan to resolve the company's unsettled work environment:

– for at least six months, no one would lose his job
– an interim committee would participate in the decision-making process
– A. Beckman would have the final authority over all decisions
– W. Shockley's contract would be extended with new assignments

In addition, Beckman brought in as a business manager Joseph Lewis, and went back on his promise to Moore to remove Shockley from his position. This foiled Moore's intention to remove Shockley, and put him in an untenable situation. He knew that he had no other choice than to look for an exit. Apart from the seven dissidents themselves, the rest of Shockley Laboratory was not aware of the dissidents' goal to remove Shockley, nor that the dissidents had no desire to seek a compromise. D. Allison described the time with the remark[12]: *"suddenly there were two groups of employees who did not communicate well with each other."* Elmer Brown, one of the Shockley Laboratory employees, wrote Shockley a letter and pointed out that the "group" was not participating in the company's goals. Shockley expected Noyce to be able to normalize this unhealthy situation, but unfortunately it did not happen.

---

[12]D. Allison in an interview with the author on May 12, 2006.

Beckman and Shockley scheduled their next meeting for July 23, 1957 and again on September 26, 1957.

On June 14, 1957 Gene Kleiner wrote a second letter to Bud Coyle "... *we believe that we could get a company into the semiconductor business within three months which would represent a considerable saving in cost and time. The initial product will be a line of silicon diffused transistor of unusual design applicable to the production of both high frequency and high power devices. It should be pointed out that the complicated techniques necessary for producing these semiconductors have already been worked out in detail by this group of people, and are not restricted by any obligation to the present organization ...*"

Shockley, in his naiveté, did not learn anything from the meeting with Beckman and Noyce. He was relieved by Noyce's statement that the group could work with Shockley. Despite warnings that the "group" was doing work behind his back, he continued to share all information freely with them.

On June 26, 1957 R. M. Ryder sent the detailed process and device description of Bell Laboratories' diffused base mesa transistor M2042 to Shockley Semiconductor Laboratory. The Shockley routing contained six names, including RNN (Noyce), JH (Hoerni), JTL (Last), and GEM (Moore). Ryder's document contained complete processing conditions, information about the equipment used, and details of the device structure and its dimensions.

In those perilous times, several bizarre and never explained events took place, including several midnight wake-up calls to Shockley or the flashy "pin affair." The timing of Shockley's wake up calls coincided with the formation of the group of dissidents, and no further wake-up call was received once the group had resigned from Shockley Laboratory. Shockley did not leave any notes regarding the pin affair and in later oral testimonies members of the group cannot even agree whether the scene of the incident was a swinging door or a bulletin board.

After the meeting between Shockley and Beckman on July 23, 1957, Beckman learned that Moore was not available for any discussions and Noyce, in line with his character trait, was trying to please everybody by telling them what they wanted to hear.

In a blow to Moore's complaint, in August 1957, Horsley's team manufactured the first 72 four-layer diodes and aimed for production in September (Fig. 12.25).

On Monday, July 15, 1957 Arthur Rock contacted G. Kleiner with news that they might have a potential investor, Fairchild Camera and Instruments, and that Fairchild's Richard Hodgson wanted to interview the seven members of the group.[13]

After burning bridges, Moore's group was hoping that the Kleiner alternative would work. They received good news in the first week of the September 1957. Hayden & Stone was working with Fairchild Camera and Instrument whose president was John Carter, former vice-president of Corning Glass Company. Carter had asked

---

[13] Arthur Rock claimed that he contacted 35 companies who turned down his proposal before he found Carter at Fairchild. [A. Rock interview with J. Markoff, May 1, 2007].

Fig. 12.25   W. Shockley and Smooth Horsley with the four-layer diode

Hayden & Stone for assistance in helping Fairchild's entrance into the transistor business. Fairchild was willing to invest up to $2 million in the new venture. A. Rock's most convincing argument was that a spin-off from the Nobel Prize winner was almost certain to be a success.

Moore's group learned that Hayden & Stone had produced a 16 page document with nine articles, prepared by a legal firm, and it had to be signed by all seven defectors on September 19, 1957. Noyce was not one of them. Sheldon Roberts negotiated with Noyce and asked him to desert Shockley. When Noyce learned that the group was able to get financing, he joined the group. The group resigned from Shockley Laboratory on September 18, 1957. All eight rebels drove the next morning to San Francisco to sign the already prepared legal documents. The problem was that Noyce's name was not in the document. The seven names typed in alphabetical order were followed by a hand written addition with Noyce's name and his address.

> Agreement dated September 19, 1957 among FAIRCHILD CONTROLS CORPORATION, New York (herein called Fairchild Controls), FAIRCHILD CAMERA AND INSTRUMENT CORPORATION, a Delaware Corporation (being called Fairchild Camera); PARKHURST & CO with principal place of business New York, N.Y. (herein called Parkhurst); JULIUS BLANK, VICTOR H. GRINICH, JEAN A. HOERNI, EUGENE KLEINER, JAY T. LAST, GORDON E. MOORE, C. SHELDON ROBERTS, and Robert N. Noyce (said eight persons being collectively called California Group.)

> The members of the California Group are scientists with extensive experience and know-how in the semi-conductor field and desire to associate with Fairchild Controls for purposes of conducting research and development with Fairchild Controls in the semi-conductor field and production and sale of semi-conductor products on the terms and conditions hereinafter set forth.

Exhibit B of the Agreement detailed the "*Semi-conductor project, estimation cash expenditures for first eighteen months of operation*." The financing of the California Group was $1.4 million for a total of 30 people of whom 22 would be technicians and junior engineers. The salaries of Shockley's dissidents almost doubled compared with what they had been getting at Shockley Laboratory.

September 18, 1957, however, did not end Shockley's troubles. On October 22, 1957, when he was preparing a reply to his patent attorney who needed additional information for the Shockley-Jones crystal growing apparatus application, Shockley found that some blueprints were missing from the folder. After some searching it was found that Knapic had all the missing documents. Shockley confronted Knapic and became suspicious. Shockley contacted the McMurray & Hamstra office and requested Knapic's background investigation. In his job application Knapic had stated that during years 1942–1946 he had been a Navy commander involved in a highly specialized operation in the South Pacific. Knapic did not wait for the result of the inquiry and resigned from Shockley Laboratory on November 29, 1957. McMurray & Hamstra's office reported to Shockley that there was no record of Knapic's Navy service.

On January 7, 1958, Dean Knapic and three other partners formed Knapic Electro-Physics, Inc. in Palo Alto, producing silicon crystals with equipment based on the Shockley Laboratory design. The Knapic company was later sold to Monsanto.

Almost all employees of Shockley Laboratory emphasize that W. Shockley had an exceptional ability to find the best talents. This claim is, of course, because Shockley selected them. Historical documents and the destiny of Shockley Laboratory proved this myth to be false; in fact Shockley hired several employees of questionable integrity.

Shockley took the defection of the "Group" very hard; he was hurt and angered. Little comfort came from A. Beckman who was now also angry, especially with Moore who had misled him. He asked his legal counsel, L. N. Duryea, to investigate the defection of Moore's group. The author found the following notes in Duryea's folder (Fig. 12.26):

The reason why A. Beckman lost confidence in G. Moore was the timing of the events. No matter what changes Beckman might have put in place in Shockley Laboratory, he would not have changed the determination of the "California Group" who, for more than six months, had been negotiating an alternative exit from Shockley Laboratory. More particulars of what actually happened came to light when two members of the "California Group" shared details of their employment at Shockley Laboratory with their girlfriends, Lorraine Tibbs and Gwen Baker. Duryea documented Lorraine's testimony: "*The group had copies of their notebook at home, as when problems come up they often stated 'I will check that at home,' and the next day they had the answer.*"

Although Duryea put together a solid legal case, Shockley refused to participate in the Beckman Instruments, Inc. legal process. A. Beckman silently welcomed

> The agreement had been negotiated over a six month period. The agreement provided that Fairchild would provide 1.38 Million Dollars over an 18 month period to a corporation to be formed by the eight individuals in exchange for a voting trust concerning the stock and an option to purchase the stock from the eight individuals. Fairchild Camera calls Fairchild Semi-Conductor its affiliate. The contact with Fairchild had been made by Eugene Kliener, one of the dissidents. His father was friendly with Hayden Stone in New York, and Arthur Rock, one of their partners, attempted to place the dissidents with a financial backer. Fairchild was finally selected.

> All manufacturers except Fairchild and Shockley use a different way of diffusion, indicating that the particular process employed by Fairchild was acquired from us.

Fig. 12.26 L. N. Duryea's notebook investigating the "California Group" (January 1958)

Shockley's position because he did not want to waste money in lengthy and expensive court litigation.

Regardless of Duryea's investigation, engineers who think for themselves can draw their own conclusions about what actually happened from the following time sequence:

January 1956: Shockley leased an empty wooden building with light and three power outlets.

February 1956: The first member of the California Group, R. Noyce, was hired. He was the only member of group with 18 months' experience at Philco.

June 1956: The last member of the California Group, J. Hoerni, was hired.

January 1957: The first silicon crystals became available at Shockley Laboratory

March 1957: Kleiner made his first contact with Hayden & Stone Co.

June 1957: R.M. Ryder sent Shockley Semiconductor Laboratory a detailed process and device description about the Bell Laboratories' diffused base mesa transistor M2042. The routing list contained the names Noyce, Hoerni, Last, Moore, and others.

September 18, 1957: The California Group resigned from Shockley Laboratory

September 19, 1957: The California Group signed an agreement with Fairchild Camera and Instruments. Inc.

The most important fact is that Fairchild's first NPN device was structurally and dimensionally equivalent to Bell Laboratories M2042 transistor.

A few years later, in 1997, we learned from Gordon Moore, when asked about the first Fairchild product, that[14]: "... *it was still to do with a double–diffused silicon transistor something that Bell Laboratories had built in the laboratory.*"[15] Yet, in

---

[14] PC Magazine, Editor Interview, March 25, 1997.

[15] In the same article Moore characterized Hoerni as "... *one of the fellows at Fairchild came up with the idea for the planar transistor.*"

the article for The Herring[16] Moore wrote: *"But taking the ideas your company is pursuing, and running down the street and setting up a parallel operation is not my idea of what is right."*

G. Moore in an interview with Charlie Rose on November 14, 2005 explained why *"the element of luck"* is important for any startup business. It is better to be lucky than good in Silicon Valley and the "California Group" was lucky when they robbed Shockley Laboratory. By today's business standards Gordon Moore and his comrades would certainly face a judge.

Shockley, who was raised by his father to *"always tell the truth,"* never understood the deceitful behavior of the "California Group." He took especially hard the desertion of Noyce and Last, whom he had considered as allies. On September 20, 1957 Shockley stated that the desertion of the "California Group," which he called the "Traitorous Eight," would have no impact on Shockley Laboratory. But he knew that the impact had been catastrophic. A few years later at the court deposition[17] Shockley stated: *"I was not the optimum director of such [an] organization. I believe my handling of it led to some dissatisfaction and some of the recruits endeavored to get me completely put out of the act. I am not sure I understood the problem very well. If I had understood it better, maybe I would have avoided it."*

Shockley's contract was extended for the next three years, effective on October 7, 1957, and Maurice C. Hanafin, co-founder of Spinco, was appointed as manager of Beckman Instruments' Shockley Laboratory. The Shockley Laboratory changed its name to the Shockley Transistor Corporation in 1958, and the company became a subsidiary of Beckman Instruments, Inc.

Shockley had experienced a great deal over the past few months and had learned much about himself and others, about business, and about human affairs in general. He became much more cautious in the hiring of new employees. In a letter to the MIT Alumni Placement Bureau he wrote *"... I would be more interested in a mature, well balanced person of satisfactory ability than a younger person who is more outstanding, but might at the same time be less stable and not adapted to the industrial and development work."*

Shockley hired a new team, including the Swiss physicist Kurt Hübner, Adolf Goetzberger from Siemens, and a fresh Ph.D from Göttingen, Hans Queisser, later followed by Hans Stark, Roland Haitz, and Hugo Fellner.

On January 19, 1959 Vincent J. Sferrino from MIT Lincoln Laboratory sent a letter to W. Shockley for fulfillment of an MIT order for 3000 PNPN four-layer diodes. Sferrino designed and assembled the fully functioning semiconductor memory comprising 1040 words of length 80 bits with an access time less than 1 μsec.

---

[16] A.B. Perkins, The Accidental Entrepreneur", The Herring (https://theherring.org), December 1994.

[17] W. Shockley versus Cox Enterprises, Inc. and R. Witherspoon, U.S. District Court for the Northern District of Georgia, Atlanta Division, Civil Action, C81-1431A.

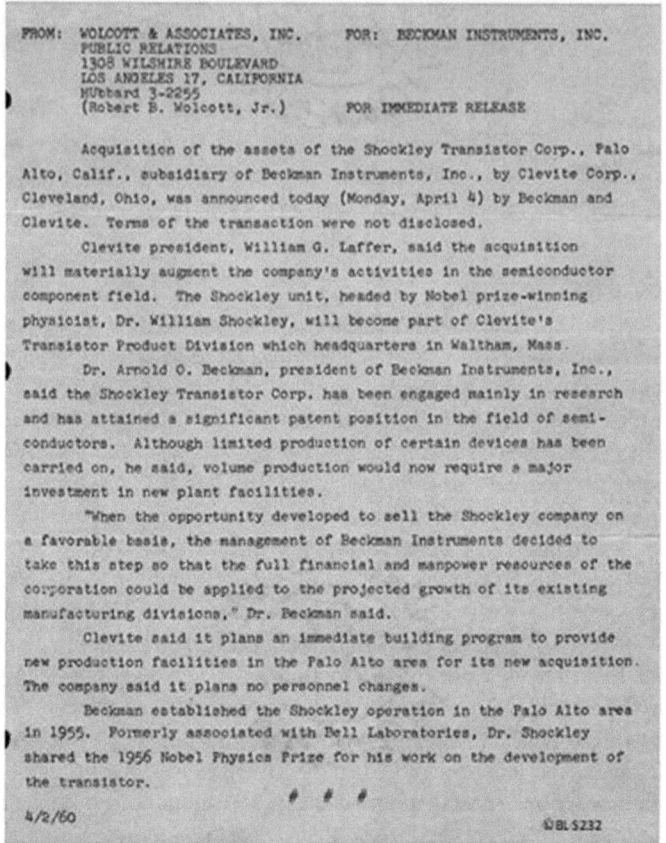

FROM: WOLCOTT & ASSOCIATES, INC.   FOR: BECKMAN INSTRUMENTS, INC.
       PUBLIC RELATIONS
       1308 WILSHIRE BOULEVARD
       LOS ANGELES 17, CALIFORNIA
       HUbbard 3-2255
       (Robert B. Wolcott, Jr.)   FOR IMMEDIATE RELEASE

       Acquisition of the assets of the Shockley Transistor Corp., Palo
Alto, Calif., subsidiary of Beckman Instruments, Inc., by Clevite Corp.,
Cleveland, Ohio, was announced today (Monday, April 4) by Beckman and
Clevite. Terms of the transaction were not disclosed.

       Clevite president, William G. Laffer, said the acquisition
will materially augment the company's activities in the semiconductor
component field. The Shockley unit, headed by Nobel prize-winning
physicist, Dr. William Shockley, will become part of Clevite's
Transistor Product Division which headquarters in Waltham, Mass.

       Dr. Arnold O. Beckman, president of Beckman Instruments, Inc.,
said the Shockley Transistor Corp. has been engaged mainly in research
and has attained a significant patent position in the field of semiconductors. Although limited production of certain devices has been
carried on, he said, volume production would now require a major
investment in new plant facilities.

       "When the opportunity developed to sell the Shockley company on
a favorable basis, the management of Beckman Instruments decided to
take this step so that the full financial and manpower resources of the
corporation could be applied to the projected growth of its existing
manufacturing divisions," Dr. Beckman said.

       Clevite said it plans an immediate building program to provide
new production facilities in the Palo Alto area for its new acquisition.
The company said it plans no personnel changes.

       Beckman established the Shockley operation in the Palo Alto area
in 1955. Formerly associated with Bell Laboratories, Dr. Shockley
shared the 1956 Nobel Physics Prize for his work on the development of
the transistor.   # # #
4/2/60   ©BI 5232

Fig. 12.27 Beckman Instruments, Inc. Press release (April 4, 1960)

The Clevite Corporation, Cleveland, Ohio, announced on April 4, 1960 that it had acquired the assets of Shockley Transistor Corp., Palo Alto, Calif., a subsidiary of Beckman Instruments, Inc. The Clevite President, W. G. Laffer, said the acquisition would substantially augment the company's activities in the semiconductor component field (Fig. 12.27).

The Press Release states: *"The Shockley unit, headed by Nobel Prize-winning physicist Dr. William Shockley, will become part of Clevite's Transistor Division in Waltham, Massachusetts."* On April 4, 1960 A. Beckman explained that the *"division has been engaged mostly in research since it was organized five years ago. It would take a large capital investment to develop needed production facilities."*

In 1961 Clevite sold Shockley Transistor to ITT, which transferred the operation to Florida. Shockley moved to Stanford University. At the same time Shockely returned to Bell Laboratories as a consultant and worked on magnetic-bubble memories. The invention of magnetic-bubble memories in 1965 by Andrew Bobeck, Richard

Fig. 12.28 Shockley Transistor, Unit of Clevite Transistor (Photo K. Hübner, 1960)

C, Sherwood, Umberto F. Gianola, and William Shockley[18] came at a time when semiconductor memories still lay in the distant future; the first semiconductor 1 k RAM, for example, did not arrive until 1970. The first magnetic-bubble memories held great promise: they were nonvolatile and they had high density. The problem was the serial access to the memory block, which resulted in long access times.

Finally, it is fair to mention that, later in their lives, at least some members of the "California Group" recognized that they should have found a better way to deal with William Shockley. Those of us who attended Austin's eulogy of Robert Noyce, and were sitting close to C. Sheldon Roberts, might have overheard Roberts' whispered remark.

One of the later Shockley employees, Kurt Hübner, told me that he asked Shockley why he chose such a very modest place for his laboratory. Shockley paused and then said: *"You do not know how hard I have been working to get where I am now."*

The last tenant of the building at 391 San Antonio Road in Mountain View before demolishing the Sears campus was International Market. The carefully formulated text on the street sign in front of the building did not say anything meaningful. Both the street sign and the original sidewalk sign were removed (Fig. 12.29).

The four real founders of Silicon Valley, W. Shockley, L. Valdes, S. Horsley, and W. Happ, have faded into history. The very first metal sidewalk plate in front of Shockley Laboratory mentioned Shockley's name, while it is barely mentioned on new Historical Sites panels. As always in history, there are plenty who are willing to take the credit for the hard work done by somebody else (Fig. 12.30).

---

[18]US Patent application # 3460116DA September 9, 1965.

Fig. 12.29   The last tenant of the building at 391 San Antonio Road in Mountain View (2008)

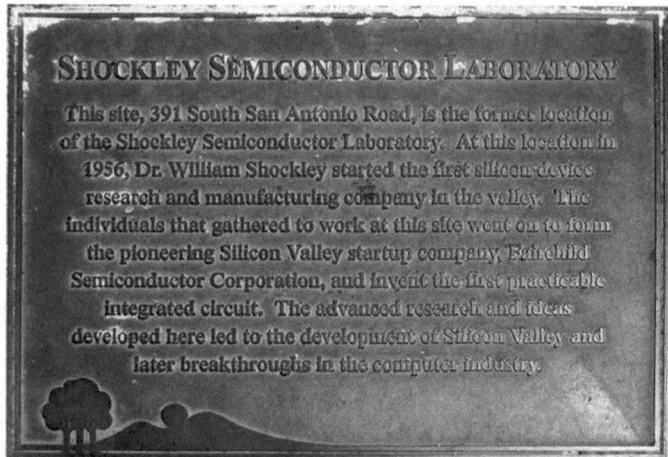

Fig. 12.30   The original sidewalk panel in front of Shockley Semiconductor Laboratory

Chapter 13
Thinking About Thinking Improves Thinking

> *"The stupidity of people comes from having an answer for every-thing. The wisdom of the novel comes from having a question for everything."*
>
> Milan Kundera

Between 1958 and 1963 Shockley served in various capacities in the Clevite Corporation and also as a lecturer at Stanford University.

An abrupt change occurred in Bill and Emmy Shockley's life on Sunday, July 23, 1961. Shockley, Emmy and Shockley's son Dick were driving on the Cabrillo Highway when a drunk driver crossed the center line and slammed into Shockley's Ford.[1] Dick did not suffer serious injuries, but Shockley and Emmy were rushed to the hospital where both were confined in heavy casts. Shockley recovered in October, Emmy in November. They both stayed on crutches for another couple of months. Shockley had dislocated his hip and had an injury to the femur.[2]

Shockley practiced self-discipline to adjust to the hospital conditions. He set up next to his bed with phone, books, and papers. His secretary Joan Altick provided a daily visit and handled all correspondence. While still in bed, Shockley wrote an article for Physical Review.[3]

In 1961 after Clevite sold Shockley Transistor to ITT. Shockley moved to Stanford University and in 1963 he was named the first Alexander M. Poniatoff Professor of Engineering and Applied Science, where he taught until his retirement in 1975.

At Stanford University Shockley taught the following classes:

EE200 "Introduction to transistor physics" (2 credits)
EE300 "Topics and methods in solid-state research" (5 credits)

---

[1] When I asked Shockley what happened, I obtained the following explanation. *"When exchange of momentum occurs during the car collision, the only way to gain or lose the momentum is for an external force to act on it. The masses were comparable but his speed was higher. Force acted on us, we were all ejected from the car."*

[2] Shockley had no head injury.

[3] W. Shockley, *"Diffusion and drift of minority carriers in semiconductors for comparable capture and screening mean free path"* Phys. Rev. Vol 125 (1965) pp. 1570–1576.

155
B. Lojek, *William Shockley: The Will to Think*, Springer Biographies,
https://doi.org/10.1007/978-3-030-65958-5_13

EE301 "Fields in crystals" (2 credits)
Seminar 92A and 92B "Mental tools for scientific thinking" (5 credits)

Quixotic Shockley, who had an unusual, intuitive understanding of physics was convinced that effective teaching methods could increase scientific thinking. While at Bell Laboratories he spent a considerable effort with the Bell Laboratories Educational Department to develop a course to teach creativity and scientific thinking. Unfortunately, to this end—alas, no positive result.

Shockley objected to the classical approach of step by step teaching at colleges when one subject is taken up, pushed through, and then after passing an examination dropped to take up another subject. Shockley's argument was that a great deal can be learned and excellent exam results be achieved, but in this manner most of the subject is rapidly forgotten. Shockley explained: *"To understand a matter thoroughly, so as really to have a lasting benefit from it, and not merely make a good showing in examination process, requires several years' familiarity. Therefore, any subject that is not kept up during the whole college course might just as well be dropped altogether and the time spent therein saved."*

Shockley advocated that theoretical courses should be accompanied by laboratory work. He frequently cited Charles Steinmetz' view[4]: *"The experiences that stay in our memory are those that are dramatic, or are exaggerated, or are unusual in some other way. Teaching experiments have shown that this principle can provide us with a useful scientific tool."*

Shockley's classes were not easy but if you were well prepared and asked questions you would learn what you could not learn from textbooks. Walter B. Hewlett, the eldest son of HP's co-founder, wrote the following in the class evaluation:

My purpose in writing this paper is to describe briefly the EE 301 seminar of the fall quarter of 1968 and to reflect on the insights it has given me into the process of research and into my own limitations within this process.

I begin with a brief description of the seminar. We (two people taking the course for credit and some auditors) met twice a week for 2 hours. The topic of the seminar was to be the study of E, B, H, and D fields. At the first meeting, Dr. Shockley explained the format. Each meeting was to be a discussion period of any material which we found interesting. The emphasis was to be on individual discovery. Dr. Shockley would interrupt occasionally to introduce self analysis, "thinking about thinking," in order that we might gain insight into our own thinking processes. Thus

the seminar was a study on two different levels. We were studying a subject, and at the same time we were studying ourselves. Lecture courses in physics are often not taught this way. All the fiddling is ironed out, so that only the polished finished product shows. Also, so much material is covered that the hurried student has time only to learn to work problems and not time to fiddle and ask questions for himself. In a sense, such smoothed out teaching stiffles questions and therefore stiffles research by presenting the subject with all the rough points covered up instead of magnified, so as to promote questioning.

I can say that before this seminar, my understanding of magnetization was very superficial, and that with quarters work, I now begin to understand the elements of the problem.

Shockley said that his first experience with unscientific thinking was when he was marked wrong on the solution of a physics problem when he was at Los Angeles Coaching School. The problem was to calculate the propelling force required to keep a boat in motion. A man was pulling each oar handle with a force of 25 pounds. The distance from oar handle to oarlock was 2 feet and that from oarlock to blade was 5

---

[4]Ch. P. Steinmetz, 1902 AIEE Presidential Address.

feet. The teacher insisted that the only correct solution of the problem was the only solution he demonstrated at the board.

The teacher regarded the oar blade as a fulcrum. The lever arm on which the man's force is pulling is then 7 feet. The force F exerted on the oar by the oarlock from the law of levers gives $25 \times 7 = F \times 5$, so $F = 35$. This would give 70 pounds pushing the boat forward, but the man pushes back on his seat with 50 pounds, so the net force on the boat is $70 - 50 = 20$ pounds (Fig. 13.2).

Shockley calculated the force using the oarlock as fulcrum, and hence found $25 \times 2 = F \times 5$, so that $F = 10$ pounds. The push on the blades is then the force pushing the boat-oars-man system forward and this force is 20 pounds.

This solution was not acceptable to the teacher because it did not calculate the actual force on the boat. Shockley and the teacher argued over this. However, Shockley's grade was not changed, and Shockley always remembered the solution of the rowboat problem as a dramatic event.

Shockley realized that this type of problem had a common feature—some overall principle permitted an answer and that was always true no matter what the details were. The power of these principles *"seemed to me to be among the more appealing esthetic aspects of theoretical sciences."* Can this combination of elegance and utility be taught? Can these methods be made more useful to science and engineering students? In the summer of 1963 Shockley asked several volunteer students to take up his pedagogical challenge. These tests resulted in Shockley's "Creative Search Pattern (CSP)" methodology (Fig. 13.3).[5,6]

Shockley formulated three important scientific tools:

1. Search thinking tools
2. Attribute—Comparison Operation—Result (ACOR)
3. Qualified law form for a result.

CSP is similar in form to various accounts of creativity and the creative process. Shockley's 1963 tests confirmed that most students could carry out some valuable and constructive analysis of their own thought process. But he added *"don't expect to avoid mistakes; acquire familiarity and get better hunches by taking action and doing something to correct your mistakes before you spend a lot of money on development."*

In 1965, the editors of Charles E. Merrill Books, Inc. implemented a new conception of physical science textbooks and started publishing the *Merrill Physical Science Series* of paperback college textbooks with copyright held not by the publisher but by the authors. The goal of the series was to present clear, concise, and correct scientific communication.

---

[5]W. Shockley, "Proposed important mental tools for scientific thinking at the high school level", Science, Vol. 140 (1963) p. 394.

[6]W. Shockley, F.J. McDonald, Report on Contract OE-4-10-216. Project S-090. U.S. Dept. of Health, Education and Welfare.

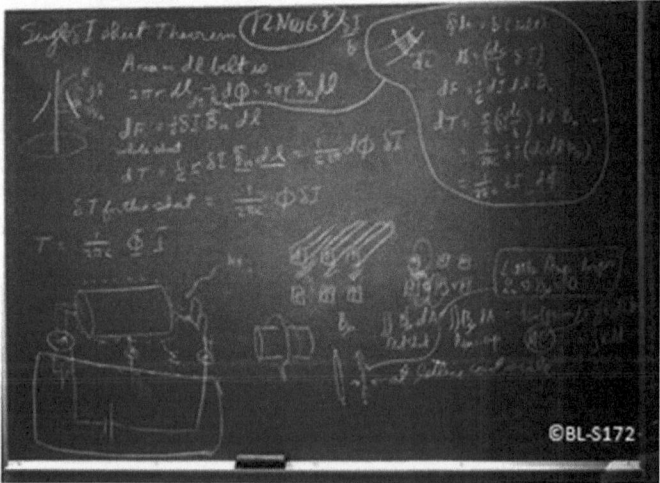

Fig. 13.1 Outline of Shockley's classes and his typical blackboard

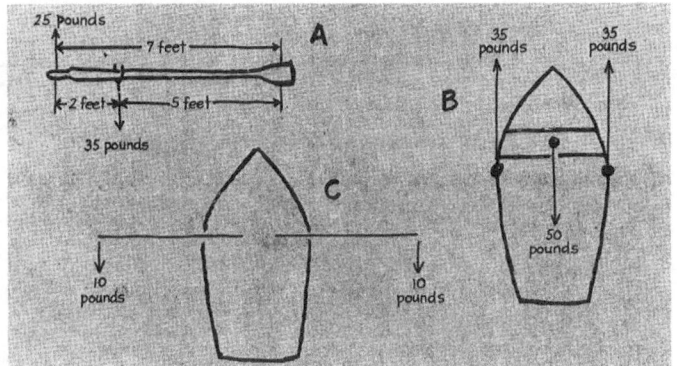

Fig. 13.2 The Los Angeles Coaching School's row boat problem (1925)

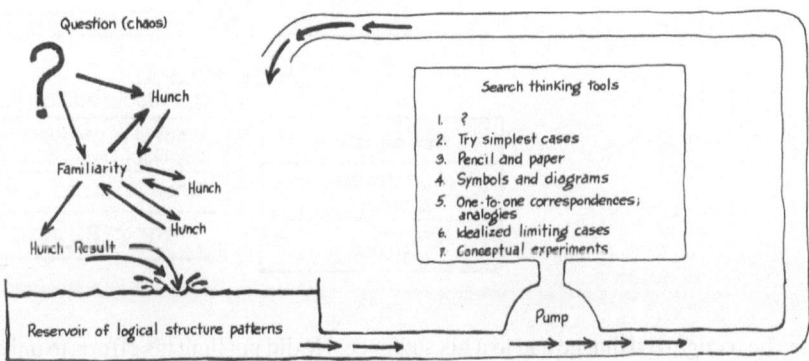

Fig. 13.3 The creative search pattern (CSP) and seven search thinking tools

San Jose State College professor, Walter A. Gong teamed up with W. Shockley in 1966 and published the textbook "Mechanics"[7] based on Shockley's four basic science-thinking tools (Fig. 13.4).

Unlike the conventional dictum of scientific method which asks for "problem definition," Creative Search Pattern starts with a question. Shockley argued that it is not always possible to define the problem at the start. Using the search-thinking tools might generate a correspondence between the problem and previously learned examples of somewhat similar problems. When a good analogy between key features of a known problem is established, a good "hunch" might transfer the old experience into a new situation.

Shockley had an unusual ability to recognize whether students were following his presentation or not. With his crisp voice he immediately rephrased his statements

---

[7]W. Shockley, W. A. Gong "*Mechanics*", Charles E. Merrill Books, Inc. Columbus, OH, 1966.

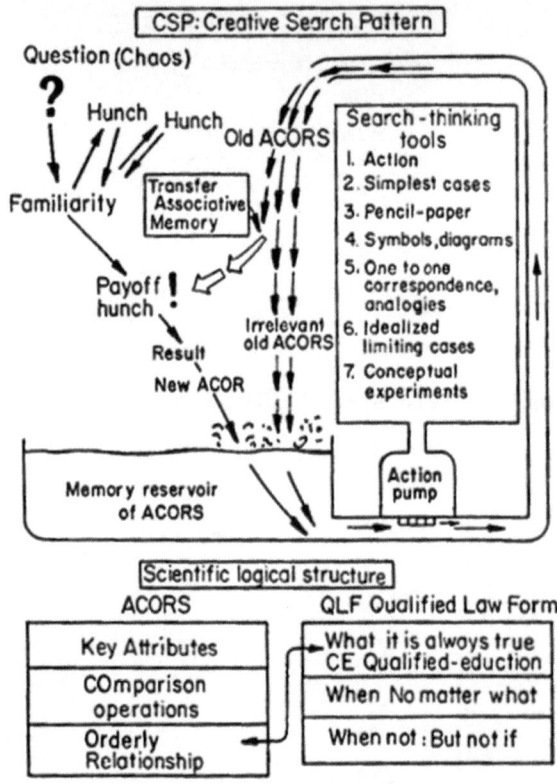

when he recognized that he had lost his audience. He did not limit his efforts to university lectures. In 1969, with a government grant, he developed the methodology of improved education for disadvantaged minority groups. The proposal, with preliminary results, was presented to the Office of Education at the West Coast Conference on Small Group Research in Los Angeles. The work was done in collaboration with Lubica Radulovic of Skyline College in San Bruno (Fig. 13.6).

In 1971 at San Jose State College Shockley lectured an extension course on child growth and development. Here, Shockley warned *"that attempts to create equality entirely by changing environment may be doomed to failure."* (Fig. 13.7).

In spite of Shockley's effort to teach scientific thinking, he gradually realized that his science-thinking tools were critically dependent on having "hunches" and that not everyone would be able to come up with such a hunch. Shockley stated *"some human beings are destined to lead unsuccessful lives no matter how hard they try … there is far more cruelty involved in being idealistic and wishful instead of finding out what the facts are."*

Only a few of his Bell Laboratories, Stanford, and San Jose State College students demonstrated an increased efficiency in their scientific productivity by adopting

Fig. 13.5 W. Shockley lecturing "Thinking about Thinking" in Japan

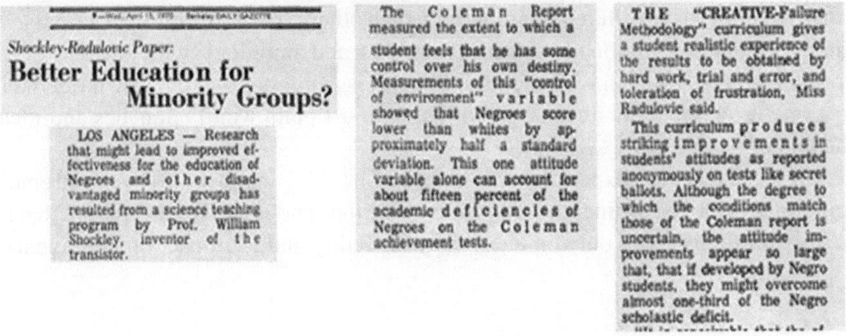

Shockley-Radulovic Paper:

Better Education for Minority Groups?

LOS ANGELES — Research that might lead to improved effectiveness for the education of Negroes and other disadvantaged minority groups has resulted from a science teaching program by Prof. William Shockley, inventor of the transistor.

The Coleman Report measured the extent to which a student feels that he has some control over his own destiny. Measurements of this "control of environment" variable showed that Negroes score lower than whites by approximately half a standard deviation. This one attitude variable alone can account for about fifteen percent of the academic deficiencies of Negroes on the Coleman achievement tests.

THE "CREATIVE-Failure Methodology" curriculum gives a student realistic experience of the results to be obtained by hard work, trial and error, and toleration of frustration, Miss Radulovic said.

This curriculum produces striking improvements in students' attitudes as reported anonymously on tests like secret ballots. Although the degree to which the conditions match those of the Coleman report is uncertain, the attitude improvements appear so large that, that if developed by Negro students, they might overcome almost one-third of the Negro scholastic deficit.

Fig. 13.6 "Improved effectiveness of the education of Negroes and other disadvantaged minority groups"

Fig. 13.7 W. Shockley
during tests performed at San
Jose State College

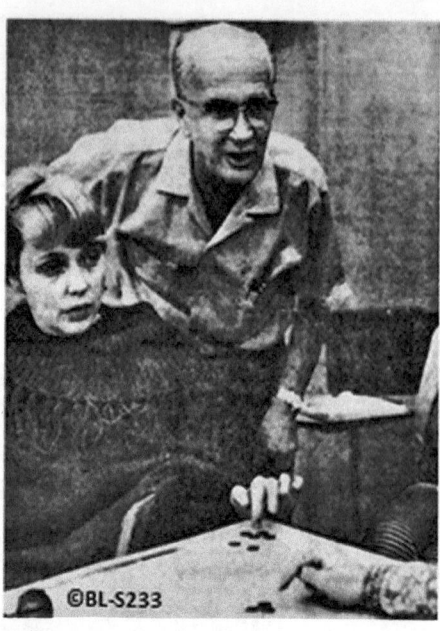

©BL-S233

Shockley's science teaching. Shockley realized that believing that everyone has the capacity to be just as creative as the next person is as ludicrous as believing that everyone has the capacity to be just as intelligent as the next person. Creative people are different from other people. And everything we know about them suggests that they are creative because they are different, not that they are different because they are creative. It is a vital distinction. Gordon Torr[8] cites a scientific study confirming that creative people have different brain activity than others. They are childlike, impulsive, fantasy oriented, emotionally sensitive, anxious, and ambitious, with lower levels of cortical arousal, which means their thinking is less inhibited and they are more likely to come up with "more absurd, dreamlike and just plain weird" hunches and ideas than other people.

All these teaching experiments convinced Shockley that heredity was extremely important in the determination of cognitive abilities, and it was for this reason that he focused in future work only on the study of heredity and its impact on intelligence.

---

[8]G. Torr, *"Managing creative people"*, The J. of Product Innovation Management, Vol. 26 (2009), pp. 467–468.

Chapter 14
Moral Philosophy as Applied Science

"The wishful thinking that causes the illusion of plasticity of intelligence is eloquently described as 'No one knows' or 'There is no way to tell'."

W. Shockley.
Lecture at UC Medical School San Francisco, Nov. 29, 1967.

Over the past years Shockley had learned much about human behavior in general. Most of it, including the deceptions of former employees, had lowered his opinion of human affairs. While teaching courses at Stanford University, he turned his attention to social problems. He was concerned with the fact that social problems in society were not being solved and did not correlate with the fast advances in technology and science. He argued that, when we reflect on the numerous lives lost during World War II, we think of weapons as responsible for these deaths. Humanity has often escaped destruction because of clear scientific thinking in the face of danger. Shockley always drew the conclusion that one of the most important distinctions in war is between those who fight for their country, and those who do not fight. Shockley witnessed the human tragedy and horror of the Bengal famine as a consequence of World War II. Such humanitarian concerns triggered Shockley's interest in social issues.

Shockley's concerns for humanity turned into his own personal catastrophe. What is sad is that he drove himself to his own destruction with theories which eventually made him an academic pariah. The PBS program "Transistorized!" broadcast in 1999 concluded: *"Shockley was vilified, ridiculed, humiliated, and eventually forgotten"*.[1] Another web site[2] offered the following characterization:

- *But Shockley the brilliant scientist had another side—white supremacist and eugenics proponent. He was convinced that race-based IQ differences existed and spent most of his career after the 1960s promoting his racist theories and a high IQ sperm bank.*

---

[1] http://www.pbs.org/transistor/album1/shockley/shockley3.html.

[2] https://www.sfgate.com/tech/article/Silicon-Valley-Shockley-racist-semiconductor-lab-131642 28.php.

B. Lojek, *William Shockley: The Will to Think*, Springer Biographies,
https://doi.org/10.1007/978-3-030-65958-5_14

163

- *Shockley left to take an engineering professorship at Stanford, where he became obsessed with racial genetics despite having no training in the field. He began espousing his radical beliefs in public forums.*

Yet another site[3] provides this characterization:

- *But Shockley's managerial style would prove severe and suspicious. As one friend would observe, Shockley had a kind of reverse charisma—when he walked into a room, you instantly took a disliking to him.*
- *In 1965, he preached at a conference that the human race was threatened by a "genetic deterioration" and the attention he received from a U.S. News and World Report interview a month later further fueled his desire for verbal combat. His views became increasingly controversial, as he asserted that darker races were mentally inferior to whites and that ghetto blacks were "downbreeding" humanity.*

I was alarmed by the claim about "reverse charisma" and questioned its accuracy. In the exact sciences, accuracy is a virtue. One sign of scientific progress is increasing precision, often, but not necessarily, expressed mathematically. Of course, there are significant differences between the exact sciences and social sciences. In contrast to the exact sciences, it is difficult to employ a highly scientific methodology in the social sciences. We have achieved significant progress in the exact sciences, but the debate around social studies consists mainly of opinion-led discussions with vague or ambiguous definitions of concepts which allow a great deal of leeway for interpretation.

How exact were the PBS and similar statements? What exactly were Shockley's "radical beliefs"? Rather than arguing about whether the PBS and other assertions were accurate, and rather than seeking out the source of such claims, it will be more constructive to provide the reader with a summary supported by archived documents.

Two major events triggered Shockley's interest in heredity and intelligence. One was an incident in which blinding acid from a baby bottle was thrown in the eyes of San Francisco delicatessen proprietor Harry Goldman in the Fillmore (a music venue) by teenager Rudy Hoskins. This was a big story[4] in 1963 and was probably more influential than any other single cause in initiating Shockley's concern with possible dysgenics[5] effects on modern society.

Rudy Hoskins, nicknamed "The Brute", had an IQ of 60–65 and was one of 17 illegitimate children of a woman reported to have an IQ of 55, who could remember the names of only nine of her children.

Shockley's interest in intelligence was triggered by the director of research at the Educational Testing Service, Western Office in Berkeley, who mailed Shockley an

---

[3] http://www.paloaltohistory.org/william-shockley.php.

[4] San Francisco District Attorney's Office trial transcript and clippings. San Francisco History Center, SFH 93, Box # 8.

[5] Shockley's definition of dysgenics: *"Dysgenics—the study of mechanisms adverse to human genetics quality."*

article by J. P. Guilford.[6] At that time Shockley was working on the "Shockley-Gong Physical Science Thinking Diagram." He immediately checked all the references in the article. Disturbed by the consequences of Hoskin's heinous crime, Shockley began to study the details of heritability and its relation to intelligence.

The problem had a long history. Many years before, the American psychologist and education researcher Lewis Terman had claimed connections between intellectual ability and race. He was a proponent of eugenics, a social movement aiming to improve the human "breed" by perpetuating certain inherited traits and eliminating others. He pushed for the forced sterilization of thousands of "feebleminded."

In 1916, Terman wrote: *"High-grade or border-line deficiency ... is very, very common among Spanish-Indian and Mexican families of the Southwest and also among Negroes. Their dullness seems to be racial, or at least inherent in the family stocks from which they come ... Children of this group should be segregated into separate classes ... They cannot master abstractions but they can often be made into efficient workers ... from a eugenic point of view they constitute a grave problem because of their unusually prolific breeding."*

Terman endorsed the 1922 circular of the International Commission on Eugenics calling for a movement *"to stem the tide of threatened racial degeneracy."* As a proponent of eugenics, aiming to perpetuate certain inherited traits and eliminate others, Terman never publicly recanted his beliefs. Because his view in his day was not much different from the views of most white people of his generation, he is remembered today as the first really big name in Stanford and his views on eugenics are now reduced to footnotes.

W. Shockley brought the issue to the fore again forty years later, at a time when social changes had produced a quite different society. The problem of social inequality persisted, but almost no one was willing to address the issue. One of those brave enough to do so was Willard Wirtz, the Secretary of Labor in the Johnson administration, who developed the program "War on Poverty." In 1964, Wirtz stated[7]: *"There is a strong indication that a disproportional number of unemployed come from large families, but we do not pursue evidence that would permit establishing this as a fact or evaluating its significance."* Shockley contacted secretary Wirtz who, in a reply to a letter from Shockley, wrote[8]: *"I hoped that this statement would encourage someone to ferret out facts."*

In his research Shockley always used government data. The US Army Alpha and Beta Tests screened approximately 1.75 million draftees in World War I in an attempt to evaluate the intellectual and emotional temperament of soldiers. Results were used to determine how capable a soldier was of serving in the armed forces and to identify which job classification or leadership position they were most suitable for. Starting in the early 1900s, the U.S. education system also began using IQ tests to identify "gifted and talented" students, as well as those with special needs who required

---

[6]J.P. Guilford, *"Intelligence: 1965 model"*, American Psychologist, Vol. 21, (1966) pp. 20–26.

[7]*"The Negro Family, The case for national action"*, Office of Policy Planning and Research, U.S. Dept. of Labor, 1965.

[8]W. Wirtz, letter to W. Shockley, February 4, 1965.

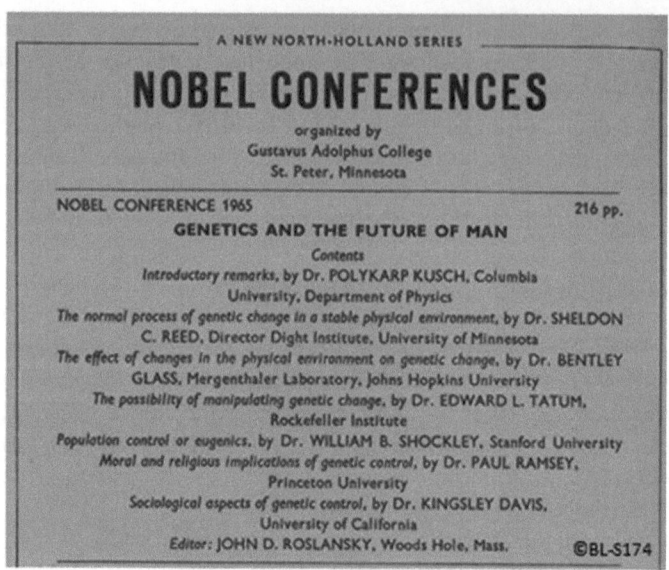

Fig. 14.1 Program of the 1965 Nobel conference at Gustavus Adolphus College

additional educational intervention and different academic environments. Because of Shockley's war involvement, he had access to the government data. Shockley analyzed all the available data, and the results of this analysis led him to conclude that there was indeed a relationship between heredity and intelligence.

After winning the Nobel Prize in physics and starting his own company, Shockley was considered a leading international figure within the scientific community. However, things then took a controversial turn in his career; in 1965, Shockley attended a Nobel conference titled "Genetics and the Future of Man" organized by Gustavus Adolphus College in Saint Peter, Minnesota (Fig. 14.1). On January 7, 1965 Shockley presented his speech: "Population control or eugenics". Shockley said *"An exponential explosion of technological advances has largely eliminated 'survival of the fittest' as a control mechanism in man's evolution."* Although almost all of Shockley's presentation deals mainly with the population explosion and the risks it involves, he makes the following remark in one part of the presentation: *"Many thoughtful people are now concerned about the possible genetic deterioration due to selective multiplication of less gifted members of society through extremely large families or high rates of illegitimacy"*.

The most challenging idea in Shockley's presentation was a voluntary sterilization "Thinking Exercise"[9]: *"I have only one positive suggestion to make, a proposal which now seems so farfetched that I find it creates only amusement when I propose it. A system of marketable licenses to have children is the only one which will combine the minimum of social control necessary to this problem with a maximum individual*

---

[9]Shockley never advocated any compulsory sterilization program.

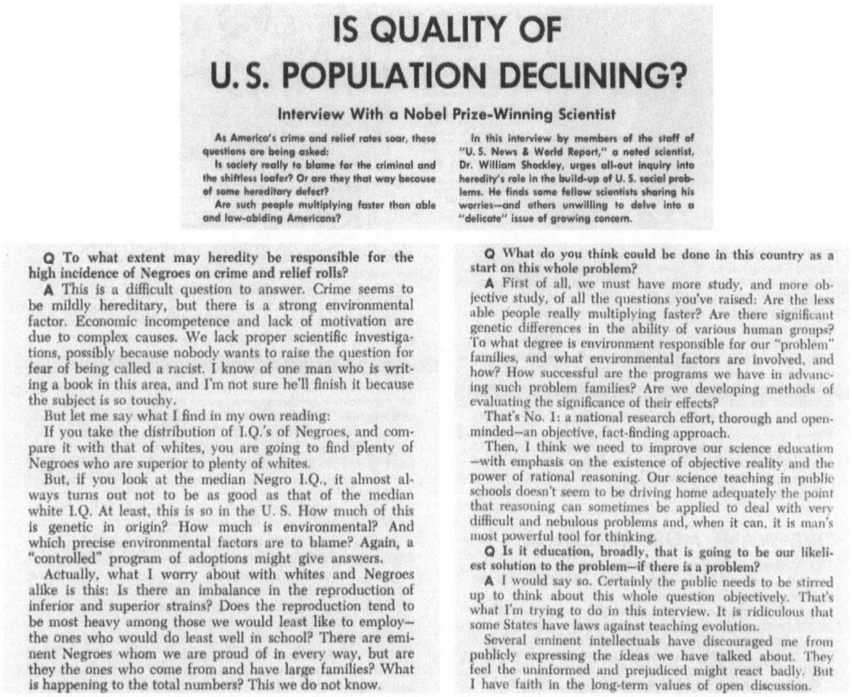

IS QUALITY OF U. S. POPULATION DECLINING?

Interview With a Nobel Prize-Winning Scientist

As America's crime and relief rates soar, these questions are being asked:

Is society really to blame for the criminal and the shiftless loafer? Or are they that way because of some hereditary defect?

Are such people multiplying faster than able and law-abiding Americans?

In this interview by members of the staff of "U. S. News & World Report," a noted scientist, Dr. William Shockley, urges all-out inquiry into heredity's role in the build-up of U. S. social problems. He finds some fellow scientists sharing his worries—and others unwilling to delve into a "delicate" issue of growing concern.

Q To what extent may heredity be responsible for the high incidence of Negroes on crime and relief rolls?

A This is a difficult question to answer. Crime seems to be mildly hereditary, but there is a strong environmental factor. Economic incompetence and lack of motivation are due to complex causes. We lack proper scientific investigations, possibly because nobody wants to raise the question for fear of being called a racist. I know of one man who is writing a book in this area, and I'm not sure he'll finish it because the subject is so touchy.

But let me say what I find in my own reading:

If you take the distribution of I.Q.'s of Negroes, and compare it with that of whites, you are going to find plenty of Negroes who are superior to plenty of whites.

But, if you look at the median Negro I.Q., it almost always turns out not to be as good as that of the median white I.Q. At least, this is so in the U. S. How much of this is genetic in origin? How much is environmental? And which precise environmental factors are to blame? Again, a "controlled" program of adoptions might give answers.

Actually, what I worry about with whites and Negroes alike is this: Is there an imbalance in the reproduction of inferior and superior strains? Does the reproduction tend to be most heavy among those we would least like to employ—the ones who would do least well in school? There are eminent Negroes whom we are proud of in every way, but are they the ones who come from and have large families? What is happening to the total numbers? This we do not know.

Q What do you think could be done in this country as a start on this whole problem?

A First of all, we must have more study, and more objective study, of all the questions you've raised: Are the less able people really multiplying faster? Are there significant genetic differences in the ability of various human groups? To what degree is environment responsible for our "problem" families, and what environmental factors are involved, and how? How successful are the programs we have in advancing such problem families? Are we developing methods of evaluating the significance of their effects?

That's No. 1: a national research effort, thorough and open-minded—an objective, fact-finding approach.

Then, I think we need to improve our science education—with emphasis on the existence of objective reality and the power of rational reasoning. Our science teaching in public schools doesn't seem to be driving home adequately the point that reasoning can sometimes be applied to deal with very difficult and nebulous problems and, when it can, it is man's most powerful tool for thinking.

Q Is it education, broadly, that is going to be our likeliest solution to the problem—if there is a problem?

A I would say so. Certainly the public needs to be stirred up to think about this whole question objectively. That's what I'm trying to do in this interview. It is ridiculous that some States have laws against teaching evolution.

Several eminent intellectuals have discouraged me from publicly expressing the ideas we have talked about. They feel the uninformed and prejudiced might react badly. But I have faith in the long-term values of open discussion.

Fig. 14.2 Snippets from U.S. News and World Report (November 22, 1965, pp. 68–71)

liberty and ethical choice. Each girl on approaching maturity would be presented with a certificate which will entitle its owner to have, say, 2,2 children or whatever number would ensure a reproductive rate of one. The unit of these certificates might be the "deci-child", and accumulation of ten of these units by purchase, inheritance, or gift would permit a woman in maturity to have one child. We would then set up a market in these units in which the rich and the philoprogenitive would purchase them from the poor, the nuns, the maiden aunts, and so on."

The presentation at the Nobel Conference did not mention any specifics about the black population but attracted the attention of the media because Shockley touched some issues which Johnson's program of The Great Society and its war on poverty ignored. U.S. News and World Report requested an interview which Shockley granted. In 1965, U.S. News and World Report had more than 400,000 subscribers and Shockley assumed that his interview with the magazine would spread his message (Fig. 14.2). Shockley wished to discuss the negative demographic trends in American society, but the editorial board of the magazine was looking for a sensation—the low IQ performance of the black population.

Shockley was always very careful to present data which were indisputable. His favorite argument was: *"If what you are telling me really makes any sense, you should be able to draw a graph of it."* And this is what he always did himself in his

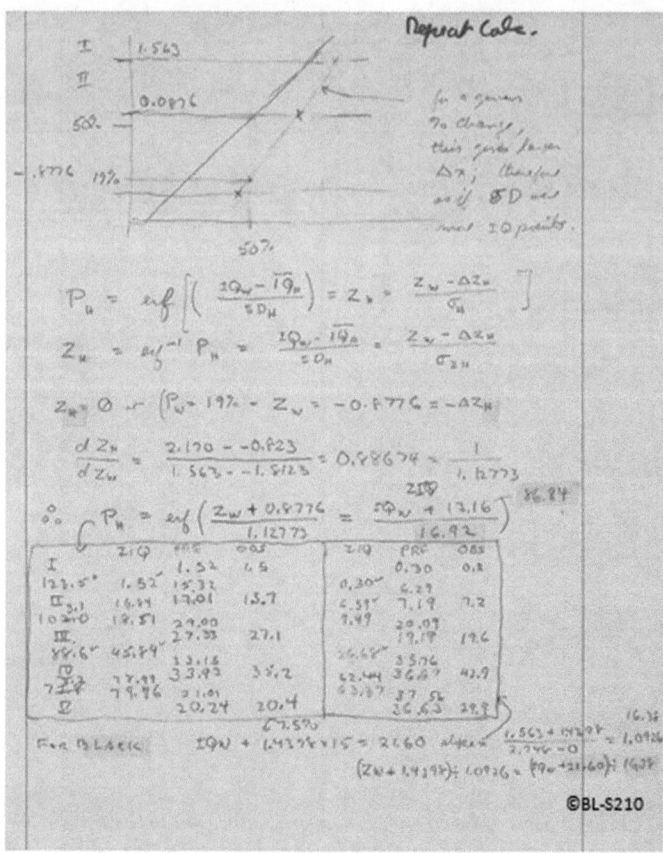

Fig. 14.3 Example of a statistical calculation for the 1965 Nobel Conference

presentations or publications. Shockley excelled in statistics and probability theory, his analyses, which he would do for any fact he presented, are error-free. Shockley would not present any chart or result unless he had checked that all was correct (Fig. 14.3). Up to this day, no one has proved that a Shockley analyses was in error. So, because Shockley's analysis was correct, some people, like Dr. Edward Scanlon, for example, claim[10] that the government data Shockley used *"are poorly controlled and administrated."* The only problem with Scanlon's argument is that all new data published by the U.S. Department of Education show the same trend as the ones Shockley used (see Ref. 6 in Chap. 1).

In addition, Shockley checked several statements before his interview with U.S. News and World Report. Dr. Richard L. Masland, Director of the National Institute of Neurological Diseases and Blindness, confirmed Shockley's findings, but in a reply to Shockley dated August 20, 1965 stated: *"…as you suggested I am writing down*

---

[10]The Charlotte Observer, Wednesday, September 8, 1971, p. 2A.

some points about your manuscript ... because of the sensitivity of these matters, I prefer very much to remain apart from any personal basis, and I am asking you not to be quoted on these issues."

Readers may judge for themselves whether the "Thinking Exercise" is or is not a morally acceptable solution. Shockley just asked people to think about it. His arguments were based on the fact that members of a technological society with low IQ are less employable and are therefore prone to dependency on others.

As a result of the U.S. News and World Report interview, Shockley learned an important lesson: dealing with communication media might be counterproductive. U.S. News and World Report did not publish all the answers Shockley provided.

On March 18, 1968 Shockley presented his "Nine Position Statement" at the University of Massachusetts in Amherst. Many press reports significantly deviate from what Shockley presented, and also from a copy he provided to the press. Both factually and grammatically, there are significant differences between the words Shockley read and those printed in the newspapers:

(1)   Assertion of lowering IQ

AP (Associated Press News): It can be proved on the basis of now available facts, the speaker said, "that an actual loss of ground ..."

UPI (United Press International): High birthrates have lowered "the negro genetic capacity of intelligence ..."

Nine Point Statement: *"Although I do not believe it has been proved I conjecture that it can be proved on the basis of now available facts that an actual loss of ground for Negro genetic potential has indeed occurred ..."*

(2)   Present Negro IQ

UPI: Shockley said the defects of Negroes in slums are hereditary ...

Nine Point Statement: *"The available facts lead me to fear that ghetto birth rates are lowering Negro hereditary potential for intelligence ..."* (Fig. 14.4).

Undeterred, Shockley issued on April 1, 1968 the following statement: *"I am prepared to suffer the consequence of such inaccuracies as appear in both AP and UPI reports in the interests of informing the public of the general nature of my worries."*

In a phone conversation with his friend Dr. Lorincz on April 20, 1968, Shockley stated:

S; let me tell you my attitude towards newspaper people. They are like graduate students and they should have lecture notes for them—they have difficulties to get things straight.
L: Misinterpreting things.
S; Not being too bright and looking for sensational things.

NINE POINT POSITION STATEMENT FOR UNIVERSITY OF MASSACHUSETTS

By Dr. W. Shockley

1. I am in favour of welfare programs in general and Head Start in particular.

2. I believe our society can and should endeavour to formulate programs so that every baby born has high probability of leading a dignified, rewarding and satisfying life.

3. Although I conjecture that some form of eugenics will be essential to achieving my second point, my recommendations regarding eugenics are restricted to the demand for objective inquiry. Eugenics is now so shunned a subject for discussion that a foundation for wise action decisions is lacking.

4. I do favour complete availability to all citizens of birth control information and supplies and complete liberalization of abortion laws.

5. My attention in the last three years has been brought to focus on the genetic potential for intelligence of the illegitimate, ghetto Negro baby for two reasons:

First, the sickness of our nation shown by the problems of racial unrest are agonizing to all responsible citizens and are obviously most acute for the disadvantaged Negro minority.

Second, the available facts show that ghetto birth rates are lowering Negro hereditary potential for intelligence so that the result may be a form of genetic enslavement that may provoke extremes of racism with resultant misery for all our citizens.

6. Although I do not believe that it has been proved I conjecture that it can be proved on the basis of now available facts that an actual loss of ground for Negro genetic potential for intelligence has indeed occurred during the last 30 years as an unforeseen by-product of the encouragement that our welfare programs have given to the least effective elements of our population to have large families; this probably occurs for white as well as black but disproportionately more for the black. Let me emphasize again that I endorse welfare programs What I urge is objective inquiry to see if my fears are justified. If my fears are justified and their recognition leads to remedial changes in welfare programs, then all citizens, again regardless of race, will benefit more from the abundance made possible by our outstanding national productivity.

7. I do believe that many American Negroes are superior to many whites. In fact my statistical studies show that Negroes achieve almost every eminent distinction that whites achieve. However, the probability on a per capita basis is about ten to one hundred times smaller and it is this probability that I fear is falling as a result of high ghetto birth rates.

8. I believe my actions in raising these questions are like those of a visitor to a sick friend who urges a thorough diagnosis, painful though the diagnosis may be, so that remedial steps may be based on objectively established facts and sound methodology. To fail to raise these unpopular questions because of fear of the resentment towards me that may ensue is an irresponsibility I am not willing to have on my conscience. I believe and hope that my determination to see that these questions are faced and answered may be the greatest contribution anyone can make to American Negro welfare for the next generation.

9. During the last thinking five minutes of my life I hope to consider that during 1968 I used my capacities to their maximum potential with the aim, as phrased in Nobel's will, of "conferring greatest benefit on mankind."

Fig. 14.4 Shockley's Nine Point Statement presented at the University of Massachusetts

Only one newspaper in the country, The Charleston News and Courier, South Carolina had the courage to call attention to Shockley's disturbing proposal and described the Nine Point Statements differently than AP or UPI. The editorial published on Thursday, March 21, 1968 reads (Fig. 14.5).

SCIENCE AND RACE

A scientist who still believes in science based on knowledge and practice rather than on theory is reopening a taboo aspect of race in the United States. He is meeting, we regret to note, with hostility among some of his fellow scientists, who ought to be above the political clouds obscuring truthful appraisal of ethnic facts.

Dr. William Shockley of Stanford University nevertheless courageously continues to talk about matters that ought to concern everyone regardless of color, creed or political persuasion. In a recent address to engineering students at Amherst, Mass., Dr. Shockley mentioned race and relief—meaning government welfare programs.

"The available facts lead me to fear," Dr. Shockley said, "that ghetto birth rate patterns are lowering Negro hereditary potential for intelligence."

The result, he said, is a form of genetic enslavement that may provoke "extremes of racism with resulting misery for all our citizens."

Dr. Shockley is a co-inventor of the transistor, one of the most important technological discoveries of our time. He is no cloistured scholar without practical sense to guide his scientific learning. He offers a line of thought entirely different from the "white racist" condemnation by the Kerner commission on riots.

Statistics prove, Dr. Shockley said, "an actual loss of ground for Negro genetic potential for intelligence" during the last 30 years—something he regards as "an unfortunate by-product of the encouragement that our welfare programs have given to the least effective elements of our population who have large families."

In laymen's language, we interpret Dr. Shockley's statement to mean that public welfare is sapping the intelligence of a large segment of the population, chiefly Negroes, and breeding inferior citizens of the future. If this is true, Negroes should be even more concerned than white people to find the truth and apply effective remedies. Dr. Shockley has opened important channels of thought that ought not to be ignored or dismissed as "racism."

Fig. 14.5 The Charleston News and Courier, March 21, 1968

Shockley believed in honest and unbiased scholarly research, but like almost no one else, his courage was unwavering even when the facts might be politically inconvenient. Seeking a sensational story, the media avoided discussion of data that Shockley presented, but instead called his ideas racist and compared him to a Nazi. Shockley's frequently emphasized argument that *"science knows no racial prejudice in the strict sense. Whites were included equally with blacks in my proposal for financial incentives to reduce the procreation of those of extremely low intelligence, since both the whites and the black population was threatened by the same dysgenics process"* was deliberately distorted or left out from unfavorable media coverage. From then on, Shockley recorded all conversations with news reporters.

The forum in which Shockley most vigorously pressed his case for research was the National Academy of Sciences. The idealistic Shockley believed that scientists would better understand his concerns and would be willing to think about his proposal. As a member of the Academy he had the right to propose research projects. On five occasions between April 1966 and April 1973 Shockley presented research papers about human quality problems to the Academy (Fig. 14.6).

W. Shockley began his presentation titled *"A 'try simplest cases' approach to the heredity-poverty-crime problem"* before the National Academy of Sciences on April 26, 1967 with the following statement: *"During the last five minutes of my life, I do not want to look back to 1967 and think that I failed at this meeting to present conclusions and raise questions that I know many of my intellectual colleagues believe are better left unsaid. I believe that the subjects presented in this afternoon's program are ones that a democratic society in the interests of its own preservation must thoughtfully consider and objectively discuss both privately and publicly. I cannot in good conscience fail to discuss these subjects simply for fear of the condemnation that may*

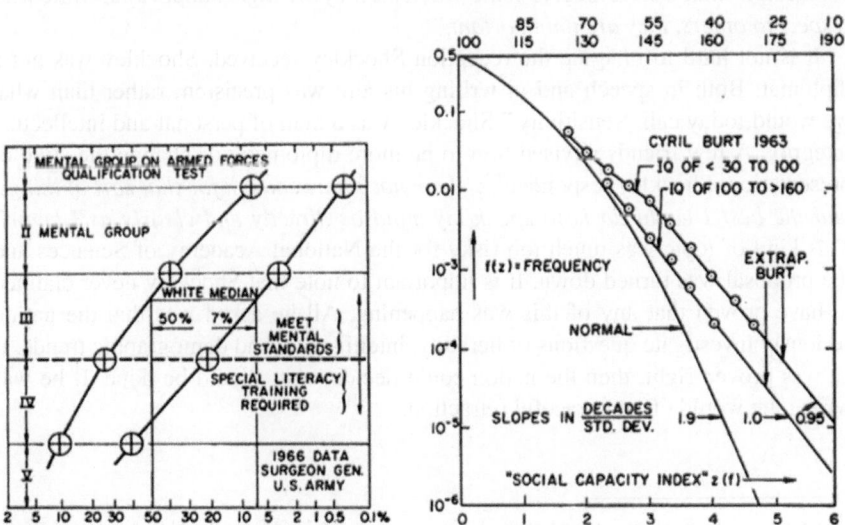

Fig. 14.6 Plot of IQ calculated from Armed Forces Qualification Test

ensue." Shockley presented the statistical analysis relevant to Wirtz's large-family statement and commented on the presented plot[11]: *"at present only 7% of the Negro scores on the Armed Forces mental tests exceed or overlap the white median score. Fifty years ago, the overlap was 13%."*

He continued with the following statement: *"The result I obtained convinced me that it must be a thinking-block, rather than the difficulty of doing research, that has kept such research from being done. My call today is for vigorous attempts to establish fact, not for any form of social action."*

Shockley's recommendations were summarized in the following paragraph:

3.   *Recommendations.*—I shall now terminate discussion of the statistical analysis and present some of my personal values.   I have placed emphasis on Negro aspects for two reasons: (1) Data are available.   (2) The racial problem is nationally prominent.

Until acceptable facts are established, the truth or falsity of the predominantly genetic simplest-case model is obscured by the environment-heredity uncertainty. I deplore this uncertainty.   If environment is the main cause, the present uncertainty will inhibit our overcoming unreasonable prejudice.   If genetics is the main cause, the uncertainty will cloud public discussions and search for solutions. Furthermore, vast expenditures in our well-intentioned war on poverty may accomplish not a solution but instead create a larger problem—a situation comparable to that of providing economic aid to underdeveloped countries and at the same time disregarding the population explosion.

An Academy committee reported: *"It is impossible to state how important genetics factors are in human quality problems taken generally."* The committee drew the conclusion *"that with respect to some problems they are highly important, while with respect to others, they are unimportant."*

It is not hard to imagine the reception Shockley received. Shockley was not a diplomat. Both in speech and in writing his aim was precision, rather than what we would today call "sensitivity." Shockley was a man of personal and intellectual integrity. A few friends advised him to be more diplomatic or change the style of presentation. Shockley responded[12]: *"I am not smart enough for that sort of things, and the best I could do is to speak my mind as directly and clearly as I know."* This kind of topic was much too risky for the National Academy of Sciences and the proposal was turned down. It is important to note that Shockley never claimed to have proven that any of this was happening. All he asked was that the nation seriously investigate questions of heredity, intelligence, and demographic trends. If he was proven right, then the nation could decide what should be done. If he was wrong, he would offer a graceful retraction.

---

[11]The data of the Surgeon general of the U.S. Army are obtained from the U.S. News and World Report, October 17, 1966, p. 78.

[12]Shockley's letter to B. D. Davis dated July 15, 1970.

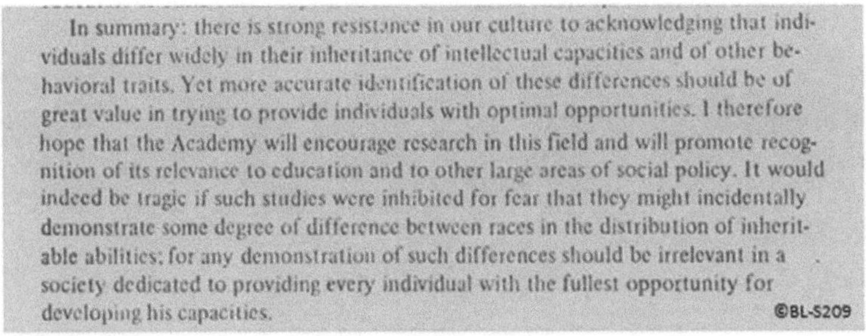

> In summary: there is strong resistance in our culture to acknowledging that indi-
> viduals differ widely in their inheritance of intellectual capacities and of other be-
> havioral traits. Yet more accurate identification of these differences should be of
> great value in trying to provide individuals with optimal opportunities. I therefore
> hope that the Academy will encourage research in this field and will promote recog-
> nition of its relevance to education and to other large areas of social policy. It would
> indeed be tragic if such studies were inhibited for fear that they might incidentally
> demonstrate some degree of difference between races in the distribution of inherit-
> able abilities; for any demonstration of such differences should be irrelevant in a
> society dedicated to providing every individual with the fullest opportunity for
> developing his capacities.   ©BL-S209

Fig. 14.7   B.D. Davis' letter to W. Shockley, dated July 2, 1970

Many of Shockley's colleagues did think he might be right, and privately encouraged him, but only a few were willing to lend him public support. B.D. Davis of Harvard Medical School wrote a letter to the president of Academy on July 2, 1970 (Fig. 14.7).

With regard to Shockley's report, the Nobel Laureate Andrew Huxley[13] wrote: *"Attempts to subordinate scientific judgement to political ends are misguided, even from a strictly political point of view. Policies based on untrue assumptions are likely to lead sooner or later to disaster."*

Shockley's biggest disappointment came from the position of his friend Frederic Seitz who was at that time the president of the Academy, and who had chosen to favor his political career. Seitz did not reply to Shockley's letter (Fig. 14.8) and their friendship died.

When the National Academy of Sciences rejected Shockley's proposal, Harvard's Harvey Brooks wrote, in a letter dated February 6, 1968: *"In one respect, I couldn't agree with you more, namely, urgency of providing the means to the poorest member of our society to prevent the conception of children with poor heredity and a guaranteed deleterious environment. I would insist that our society must do everything in its power to see that the children whose appearance in the world was unfortunately not prevented are not punished and deprived because of the sin and disability of their parents."* However, a few months later Harvey Brooks replied to Shockley with cc. F. Seitz on May 4, 1966 and explained his real position: *"I am afraid that in the present climate it is very probably the kiss of death as far as anybody listening to you objectively is concerned. If the statement about mean white and non-white IQ's did not have such touchy implications, it would probably have [remained] unnoticed".*

Shockley's proposals were rejected year after year. Eventually, he lost patience. He had repeatedly warned the Academy that *"if differential birth rates really had thrown evolution into reverse, it was a catastrophe that required immediate attention"*.[14] In

---

[13] Palo Alto Times, September 1, 1977.

[14] Honorable Ch. S. Gubser of California, *"NAS Tables Resolution"*, in Congressional Record "Proc. and Debates of the 91st Congress, First Session, Vol. 115, Part 12, July 15, 1969.

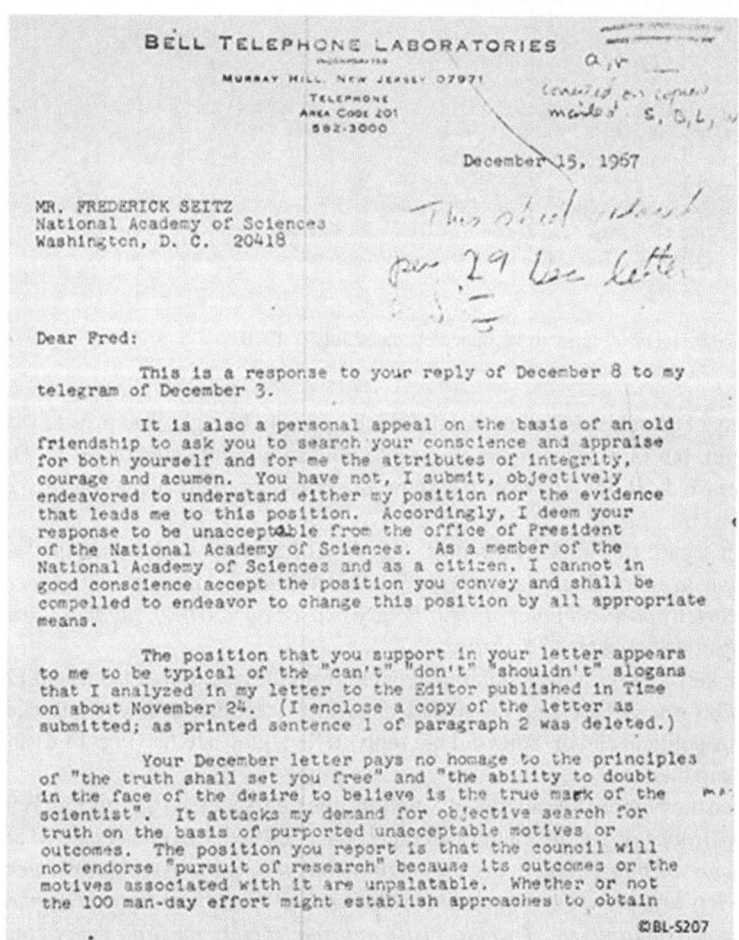

Fig. 14.8 The first page of Shockley's letter to F. Seitz, dated December 15, 1967

his last proposal in 1972, he called the Academy's inaction *"the most serious and obvious dereliction of intellectual responsibility in the history of science."*[15]

It is clear from many of Shockley's papers and presentations that, as far as he was concerned, it was the sacred duty of scientists to search for the truth no matter how painful the truth might be. He often told his fellow scientists that *"the courage to doubt in the face of the desire to believe is the true mark of the scientist"* and reminded them of the moral obligation to think.

The hippie movement in the late 1960s and the concurrent leftist movement made American colleges and universities an ideal place for spreading propaganda. The branches of the Black Student Union and similar radical militant organizations used

---

[15] With the advent of IBM *MagCard Selectric* typewriter, Shockley saved his presentations and most of his correspondence on the magnetic cards.

Shockley's presentations to spread their ideology through widespread press coverage. A major disruption occurred at Stanford University on January 18, 1972. Sixteen members of the Third World Liberation Front entered Room 127 in the McCullough Building at 11:00 AM where a scheduled quiz in a course on electrical engineering was being given by Professor Shockley. An unidentified black woman began passing copies of a statement to class members. The statement declared that professor Shockley had been *"found by Third World People to be racist."* It called his theories *"Nazi race theory"* and claimed that he *"was seeking to justify killing the future generation of black and other poor people."* The gang shouted that Shockley was *"advocating genocide of black people"* and demanded that Shockley meet Cedric X (Clark)[16] in a public debate before February 28, 1972. Shockley responded that *"we are advocating here quantum mechanics."* During the reading of the statement by an unidentified black man, Shockley corrected his pronunciation of "eugenics" and "dysgenics." When Shockley took a polaroid picture of demonstrators, the black man took his camera and also removed the cassette from the recording cassette recorder and passed it on to another group member who left the classroom with the tape and polaroid images.

After the disturbance, which lasted less than half an hour, Shockley agreed to give serious consideration to a request for a formal debate as long as certain conditions were met and asked the demonstrators to leave the classroom. The "Liberators" left the classroom shortly before the campus police arrived at 11:40 AM.

Encouraged by the university's inaction, five graduate students, who were members of an anti-imperialist movement of the Women's Union with the World Liberation Front, and a newspaper photographer from Venceremos, a political organization active in Palo Alto, disrupted Shockley's class again on February 3, 1972. The protesters urged Shockley to settle an alleged "racist discrimination" against Nigerian student Oladele. B. Ajayi. Ajayi was very upset because he had done badly on one particular test. In discussion with Shockley, Ajayi was resentful of Shockley's appraisal, when Shockley tried to explain and compare some difficulties that he himself had had as a graduate student.

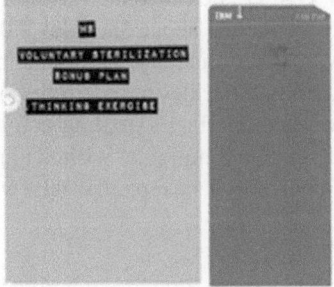

---

[16]The black activists Cedric X (Clark) was Sayed Malik Khatib, and DuBois Phillip McGee advocated the idea of the difference between African black people and Euro-American people in terms of basic culture between Africans and Europeans. Cedric X converted to Islam in 1972.

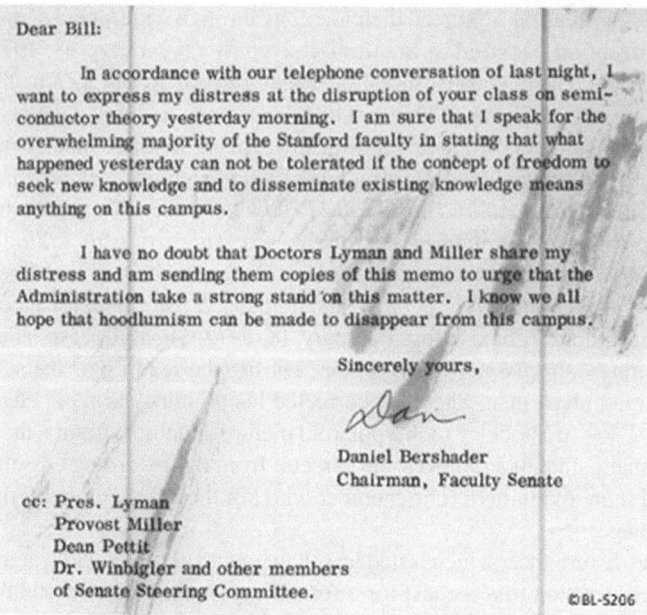

Dear Bill:

In accordance with our telephone conversation of last night, I want to express my distress at the disruption of your class on semi-conductor theory yesterday morning. I am sure that I speak for the overwhelming majority of the Stanford faculty in stating that what happened yesterday can not be tolerated if the concept of freedom to seek new knowledge and to disseminate existing knowledge means anything on this campus.

I have no doubt that Doctors Lyman and Miller share my distress and am sending them copies of this memo to urge that the Administration take a strong stand on this matter. I know we all hope that hoodlumism can be made to disappear from this campus.

Sincerely yours,

Daniel Bershader
Chairman, Faculty Senate

cc: Pres. Lyman
    Provost Miller
    Dean Pettit
    Dr. Winbigler and other members
    of Senate Steering Committee.

©BL-5206

Fig. 14.9 The university's response: *"we all hope."*

Santa Clara County Sheriff's officers arrested Betsy Elich 24, Enid Hunkeler 24, Laura King 18, Mary Cummings 20, Barbara Hyland 21, and former student John Hawkes III 20 (unemployed), all white. The photographer was charged with trespassing. They told the police officer they were from the local Ku Klux Klan and they wanted to present Shockley with a mock Ku Klux Klan award. Elich said at the sheriff's office that the disruption was intended *"as a fairly entertaining thing."*.

Later in the afternoon, Shockley met with members of the class. Five of the six class members signed a statement to the effect that Shockley had made no threat of low grade or any inappropriate comments to Ajayi.[17] Ajayi would not sign the statement, but on February 11, 1972 he called the Associate Chairman for Academic Matters, R. J. Smith, and complained about Shockley's criticism of his homework paper and said the tests were discriminatory. In an attempt to eliminate any doubt regarding impartiality in grading, Shockley asked Professor H. Heffner to grade tests from which names were deleted. Heffner's grading confirmed Shockley's original grading. When Ajayi was asked by Smith if he wished to register a complaint about Professor Shockley's treatment of him, he replied that he was not yet ready to involve the university in the affair.

---

[17] San Francisco Chronicle, February 4, 1972, p. 3.

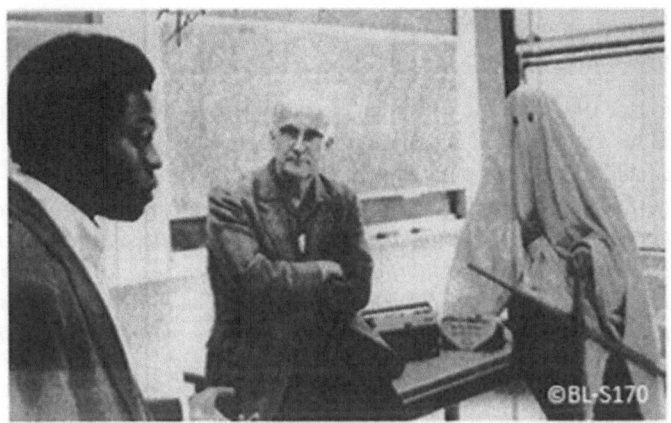

Fig. 14.10 Oladele B. Ajayi, an auditor in Shockley's class

When Ajayi failed to prove that Shockley's treatment was unfair, he dropped Shockley's and one additional course. He indicated to R.J. Smith that "*he wants to leave campus for a few days.*" Although this incident happened after the course cancelation date, the university refunded his course fee in full and offered him an individual research project.

That same afternoon about 20 members of the Liberation Front (including four Stanford students) interrupted a law school class taught by John Kaplan, who was head of the Campus Judicial Panel. Except for the university's vice provost Robert Rosenzweig, no one condemned the disruption of classes. Rosenzweig said "*the disruption of Shockley's classes and other campus disorder is a vivid threat to academic freedom.*" As expected, the student senate of Stanford University called for a committee to investigate the "*alleged racism of Professor Shockley.*"

Although the Black Student Union at different American universities played a relatively minor role in disrupting Shockley's lessons, the Stanford branch was very active. The branch organized a February 16, 1972 rally and march to the office of the Stanford University president Richard W. Lyman (Fig. 14.11). About 350 people gathered at White Plaza where Chris Fleming read a resolution demanding that Shockley be fired from Stanford University.

Alice Furumoto (far left in Fig. 14.12), former election commissioner of the Student Association, demonstrated her intellectual ability by calling Shockley a "*mother-fucking racist*" and "*racist oppressor,*" while standing right next to him.[18] After Furumoto's extempore, student Chris Yee handed Shockley a sign with the words "*One Shockley among many*". Shockley wrote on other side Voltaire's three moral postulates: "*Truth ... Concern ... Death*" and held it silently while Yee spoke. At the end Shockley said: "*I do not find any respect for the power of rational thinking*

---

[18]When I asked Shockley why he wasted time with such an aggressive and emotional mob, he replied: "*I was there simply because I would have less self-esteem if I was not there.*"

at the rally but I would be glad to debate the implications of my theories and plans with qualified representatives".

On March 6, 1972 Shockley attended a presentation by Andrew Grove at Stanford and parked his car in his usual parking space. After the presentation all four tires were nailed and the car was painted with the label "Racist Pig" (Fig. 14.14).

On May 18, 1972 Robert Rosenzweig wrote in the Stanford Daily: *"There is plenty evidence to support the conclusion that Professor Shockley has, at various times and places, been badly treated and that his rights have been seriously abused. With no more than a casual newspaper reader's knowledge of the subject, it has seemed to me that the National Academy of Sciences has treated his challenges foolishly, and*

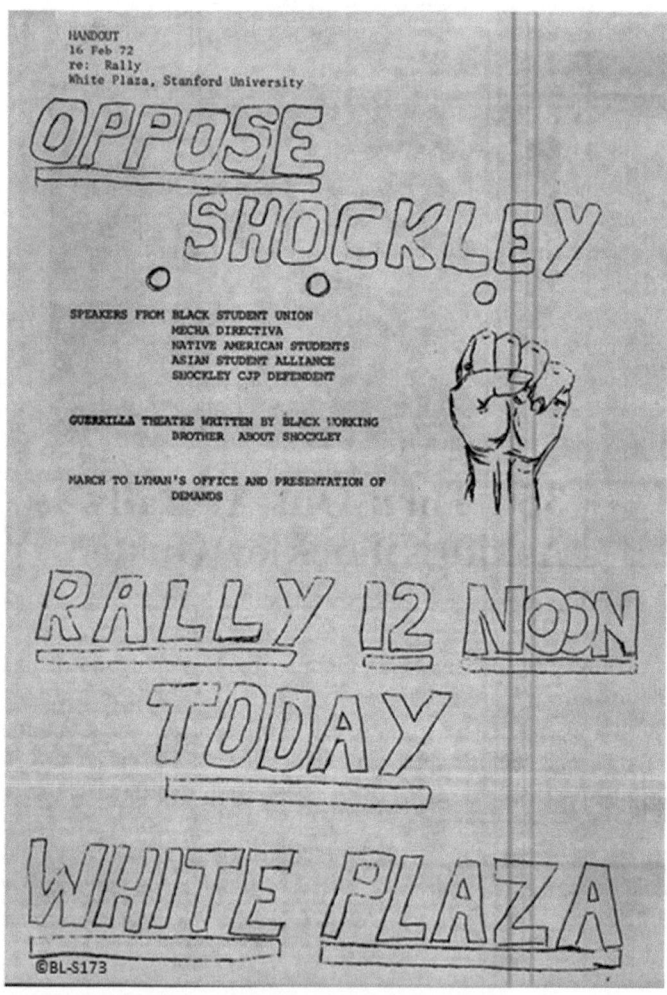

Fig. 14.11 Stanford White Plaza Rally asking for Shockley's ouster

Fig. 14.12 A member of the Black Student Union, Christopher Fleming, reads a resolution demanding. Shockley's dismissal from Stanford University

Fig. 14.13 Professor W. Shockley addressing a crowd of future Stanford intellectuals

Fig. 14.14 Shockley's car after A. Grove's presentation

perhaps unfairly. I believe too, that those people who disrupted Prof. Shockley's class at Stanford deserve to be punished."

But as always, when such dramatic situations occur, there is opportunity for personal gain. In order to boost his student ratings and popularity, Stanford professor of French, Raymond Giraud, wrote a letter to president Richard Lyman on May 31, 1972, saying: *"Dear Dick, I beg you to weigh carefully the implication of the Campus Judicial Panel's recommendation in the case of Furumoto and Lee. Harsh and excessive penalties can do irreparable damage to the lives of these young people."*

After a hearing, the Campus Judicial Panel recommended the indefinite suspension of A. Furumoto, K. Ho, and D. Lee. President R. W. Lyman adopted the recommendation. No black student involved in the class interruptions was suspended.

All three indefinitely suspended former students filed a suit against the Board of Trustees of Stanford University at the U.S. District Court for the Northern District of California.[19] Their legal team Kennedy and Rhine was paid by the World Liberation Front. The court stated that Shockley *"at no time prior to the class disruption did ever lecture on or discuss theories of eugenics and dysgenics in the course in electrical engineering."* Kennedy and Rhine claimed that the Plaintiff was deprived (1) of civil rights and privilege under color of state law,[20] and (2) the defendants sanctioned them but not the University band which disrupted class, and that the defendants prosecuted them but not others who took similar actions.

The Judge sided with Lyman, making the following statement: *"The complaint claims that Professor Shockley's writings are racist, "highly offensive" to anyone opposing racism, and "antagonize and anger" those in opposition. These claims upon analysis simply amounted to vehement disagreement with another person's exercise of his First Amendment rights, not a statement of a claim under § 1983."*

The requested debate between Shockley and Cedric X took place on January 23, 1973 in Memorial auditorium. Almost five hundred mostly white students attended the forum where Shockley faced Cedric X accompanied by DuBois Philip McGee and Stanford professor of genetics, Luigi Cavalli-Sforza. The debate was moderated by Dr. Herman Chernoff with timekeeper Dr. P. Zimbardo. Two black activists did not say anything important, apart from some propagandist clichés. When Shockley presented his data, Cavalli-Sforza suggested to the audience that on average they would not qualify as black or white and he rejected the idea of race. Cavalli-Sforza never explained the latest DNA research and the association of genetic variations with particular geographical locations which provide a genetic basis for race, but he added that *"if Shockley is my student, I would flunk him."* Each time Shockley presented his data showing the difference between blacks and whites, Cavalli-Sforza rejected

---

[19]Furumoto vs. Lyman 362 F. Supp. 1267, U.S. District Court for the Northern District of California, August 21, 1973.

[20]To act *"under color of state law"* means to act beyond the bounds of lawful authority, but in such a manner that the unlawful acts were done while the official was purporting or pretending to act in the performance of his official duties. In other words, the unlawful acts must consist of an abuse or misuse of power which is possessed by the official only because he is an official.

Fig. 14.15 Princeton students in a scientific exchange of IQ data with W. Shockley (1973)

the data with the same brash answer *"there is no such thing as race."* The great professor of genetics was speechless when Shockley argued that heritability depends on the existence of individual differences. If there were no individual differences, there would be no meaningful heritability. Cavalli-Sforza was excellent in avoiding any controversial questions, his political ramifications were more important to him than a discussion of scientific data.

The debate went on without any incident and also without any significant step in resolving differences.

The exchange of opinions between Shockley and Cavalli-Sforza changed the minds of several Stanford students. When the head of the Stanford School of Education canceled Shockley's scheduled presentation "Geneticity of IQ, Repression and Suppression of Scientific Research" on May 30, 1974, a group of students met with Ron Secrest, the head of office student personnel, and requested Shockley's presentation.

Over the years 1968–1975, Shockley made over a hundred presentations (see Appendix). The vast majority of these meetings were civil and without disturbances, even if there was disagreement between Shockley and the audience. The press, however, reported only on those meetings where large crowds gathered and boycotted the presentation from its outset. Very few knew anything about genetics and even fewer understood the statistical analyses with numbers and charts that Shockley intended to present. But Shockley was undeterred by such disturbances. He learned that the news media would not provide balanced coverage of his speeches and would pick up only what they thought were the sensational pieces. The demonstrations and disturbances at university campuses were always covered by the press, and Shockley expected at least some information about his crusade to become public.

Well-funded organizations, such as Students for a Democratic Society, had a central organizing secretariat coordinating activities on a nationwide scale. Anti-Shockley activists were transported from other universities with the objective of

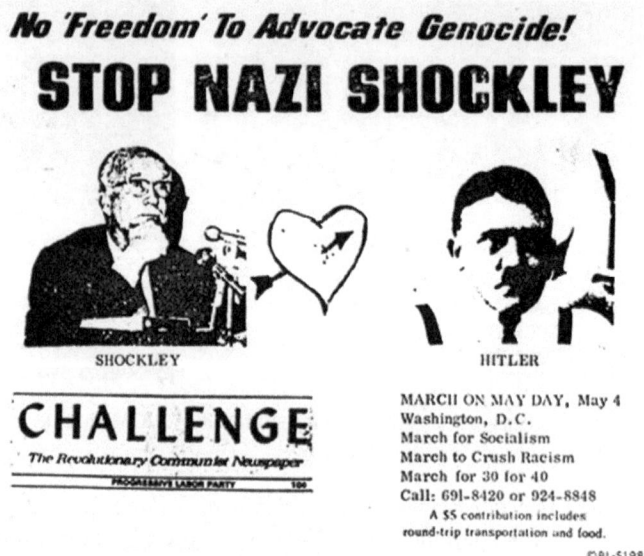

Fig. 14.16 The March for Socialism, Washington, D.C. May 4, 1974

preventing him from being heard on any campus. For example, for a contribution of $5, activists attending the "March for Socialism" in Washington, D.C. were provided with transportation and food (Fig. 14.16).

During May and September 1980 Shockley granted an interview to W. Roger Witherspoon, who had just started as health and science writer for the Atlanta Constitution newspaper. He told Shockley on June 12, 1980 that the newspaper was starting a health and science section every Thursday. They agreed to a phone interview which Shockley, with Witherspoon's approval, recorded.

R. Witherspoon was a college dropout from the University of Michigan, Ann Arbor, where he had studied two semesters of what he put in his resume as "aeronautical engineering." He was a co-founder of the association Black Journalists which eventually grew into the National Association of Black Journalists.

He characterized himself in his resume as: "*I am highly experienced at managing the analysis and communication of a wide variety of information. I am an author and educational consultant. I have evaluated, recommended and managed millions of dollars in corporate funds supporting domestic and international programs... I am an effective writer, able to present complex issues understandably and accurately.*"

Before his employment by Cox Enterprise he had changed jobs eight times in ten years. It is not clear how he qualified for the position of health and science writer without any formal education. Cox Enterprise, Inc. is a privately held company which purchased the Atlanta Constitution in 1950, with its major operating subsidiary Cox Communication. Witherspoon gave Shockley an address that was not related to Cox Enterprise.

Designer Genes By Shockley

NO ONE doubts the brilliance of William Bradford Shockley, who, along with two Bell Laboratories colleagues, invented the transistor in 1954, before they even knew how it could be used. They had the foresight to see the need for the little device, which has since revolutionized the world.

He shared a Nobel Prize in 1956 for his part in that discovery, and has spent his time since then soaking up the sun around Stanford University in California and looking for problems which may or may not have solutions. Fifteen years ago Shockley, the professional engineer and amateur geneticist, thought he had found a problem no one had had the guts to look at — the reason for the disparity of scores between whites and blacks in standard academic IQ tests.

Blacks, he said, were simply less intelligent. And they inherited this trait. And the disparities in educational opportunity, the disparities in job opportunity, the orientation of tests and testers, the effects of disparate environments had nothing to do with the fact that blacks scored 15 points or so less than whites on abstract reasoning tests — at least, not in Shockley's world. Blacks were an underclass because they were born to an underclass. Racism had nothing to do with it. Opportunity had nothing to do with it. Great Society and poverty programs could have no effect on it. Period.

The fact that the National Academy of Sciences and most ge-

Roger Witherspoon
Health and Science Writer

Nobel laureate William Shockley's genetic theories envision the manipulation of races to eliminate people deemed intellectually inferior

Fig. 14.17 The title of R. Witherspoon's article published in the Atlanta Constitution, July 31, 1980

Witherspoon's writing was published by the Atlanta Constitution on Friday, July 31, 1980 under the title "Designer Genes by Shockley" (Fig. 14.17).

Witherspoon's article completely ignored Shockley's statements provided in their interviews but provided Witherspoon's own totally outrageous interpretation of the interview, accusing Shockley outright of holding and defending Nazi policies. The article ended with the statement: *"In the end, it boils down to a man with the idea that there are too many black people around, and he is asking them to eliminate themselves."*

Shockley contacted the Atlanta Constitution with a complaint. The reply dated February 19, 1981 and signed by executive editor and vice president, James G. Minter, Jr., did not offer any remedy.

On July 28, 1981, Shockley filed[21] a $1,125.000 libel suit against the Atlanta Constitution in the Atlanta District Court, via his attorney, J. Walter Cowart of Savannah, GA. In the written statement Shockley called Witherspoon's report *"the most unwarranted derogatory presentation of my position that I can remember."* Shockley charged that certain facts pertinent to his positions on dysgenics—the reproduction of the "genetically disadvantaged"—and genetic theories relating race and intelligence should be deleted. Shockley said in his statement that the article represented him as approving of Hitler's super-race program: *"I felt that the article's widely disseminated distortions of my motivations and my humanitarian values could thwart my objective in my declining years of making a contribution to the future of humanity as great or greater than I had made in my youth by my transistor invention."*

Attorney Cowart in a letter to Shockley dated July 15, 1981 wrote: *"I am of the opinion that liberal courts (especially the Appellate Courts) have extended the*

---

[21] W. Shockley vs. Cox Enterprises, Inc. and R. Witherspoon, No. C81-1431A.

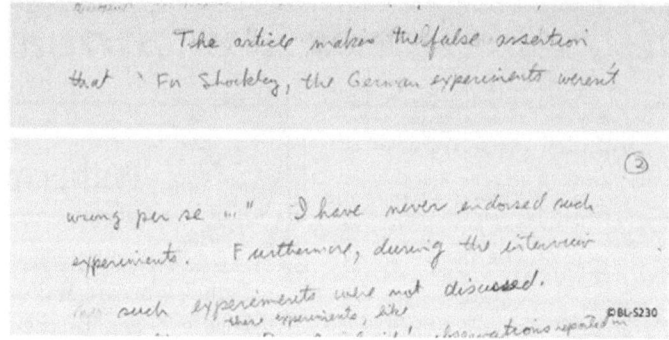

Fig. 14.18 W. Shockley's notes provided to his attorney J.W. Cowart

rules pertaining to freedom of the press to the point where it actually amounts to an abuse of ordinary citizens, all in deference to the press, and under the theory that communication in our modern society is essential to growth and education, and that private rights must be secondary to this factor. I disagree with that application." Further he provided the following explanation:

> 5. Without respite, the media has litigated the issues over and over, chipping away at the original concept of the First Amendment, injecting new applications, principles and exceptions, until now one must be able to prove actual malice on the part of the publisher in order to prevail. A base denial of equal protection of the laws occurs thereby. One who is assailed by the published half-truths, editorialized, criticized, ridiculed and abused, has no forum in which he may present his rebuttal unless he is willing to buy space in the newspaper or other media, which most often he can not do. Of course, he can still go to court.   ©BL-S232

Witherspoon was represented by David J. Bailey of Atlanta's Hansel, Post, Brandon and Dorsey and they started the deposition with Shockley's IRS Tax Return (Exhibit 366). Both Shockley and Cowart underestimated the power of Cox Enterprises, who owned the Atlanta Constitution. Shockley's 455-page deposition started with his IRS documents and covered questions about his parents, Bell Telephone Laboratories, Nobel Prize, Shockley Semiconductor, Beckman Instruments, Fairchild, and Shockley's Stanford University tenure. No question was asked regarding libel or Witherspoon's writing.

Shockley's arguments were based on the recording of all interviews and on mail exchanged between Shockley and Witherspoon (Fig. 14.18).

The transcript of the deposition from July 28, 1981 reads as follows: *"As occurred many times during the interviews, I dictated statements, pausing after every ten words or so, until Mr. Witherspoon indicated that [he] had caught up. When I remarked that his typing must be accurate, Mr. Witherspoon stated that 'I made [...] a reputation for being accurate'."*

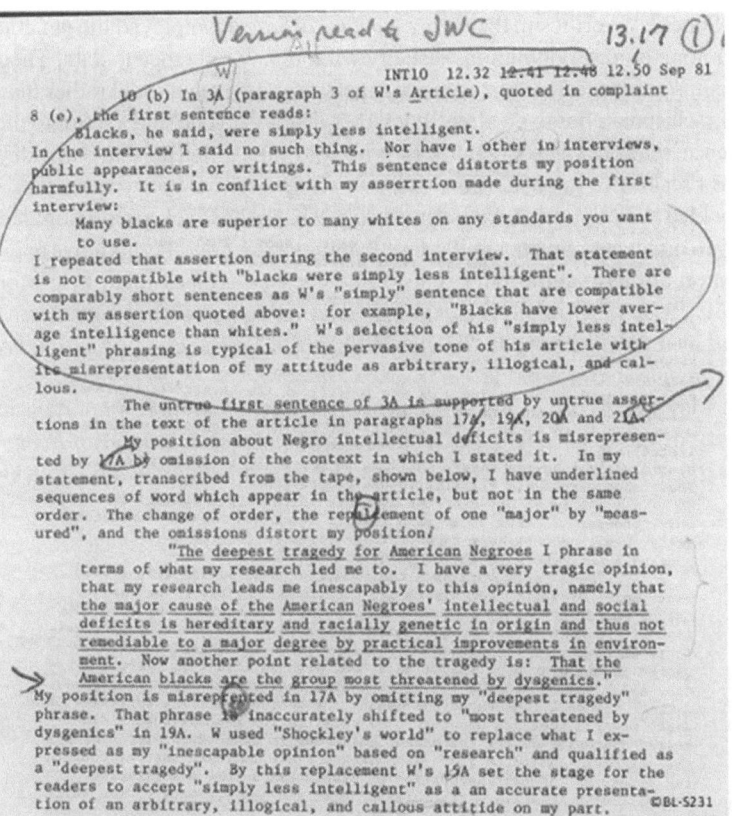

Fig. 14.19   Page from Shockley's deposition C81-1431A

Almost four years of court litigation cost Shockley over $80,000. The six-member jury, which included five white members and one black member, deliberated for about three and a half hours and vindicated Shockley that he proved the article made false statements and that it was published with reckless disregard for whether they were true or false. The jury decided in Shockley's favor.

Although Shockley's libel case was found to be valid, Judge Robert Vining awarded Shockley $1 compensation and no punitive damages.

R. Witherspoon, whose employment by the Atlanta Constitution had in the meantime been terminated, made a statement for The New York Times: "*If they had thought I was reckless or was out to get the guy, anything other than give him a fair shake, he would have gotten a heck of a lot more than a buck, and there would have been punitive damages as well.*"

All Shockley's critics in the *nature-nurture debate* downplayed the genetic differences among human populations, usually without any reference to data. Their social commentaries were often mistaken for expertise and were political rather than scientific. Hirsch spread lies even after Shockley's death when he sent a letter to the editor of Science referencing John Bardeen who told him a flat lie that Shockley had a nervous breakdown in 1951.

Shockley did not limit his dysgenics concerns to the black population; he sought equally to direct attention to similar problems among whites. Perhaps it is appropriate to mention at this point a letter sent to Shockley by a black man, Eric Morton on April 21, 1972: "*... I found it ironic that these whites who [...] vigorously oppose you and who would deny you the means to sort out the truth must be intellectually inferior people ...*"

Shockley believed that his dysgenics work was much more important than the discovery of the transistor. He explained, "*without a certain level of human intelligence, there could be no transistors or much else, for that matter.*" Shockley was

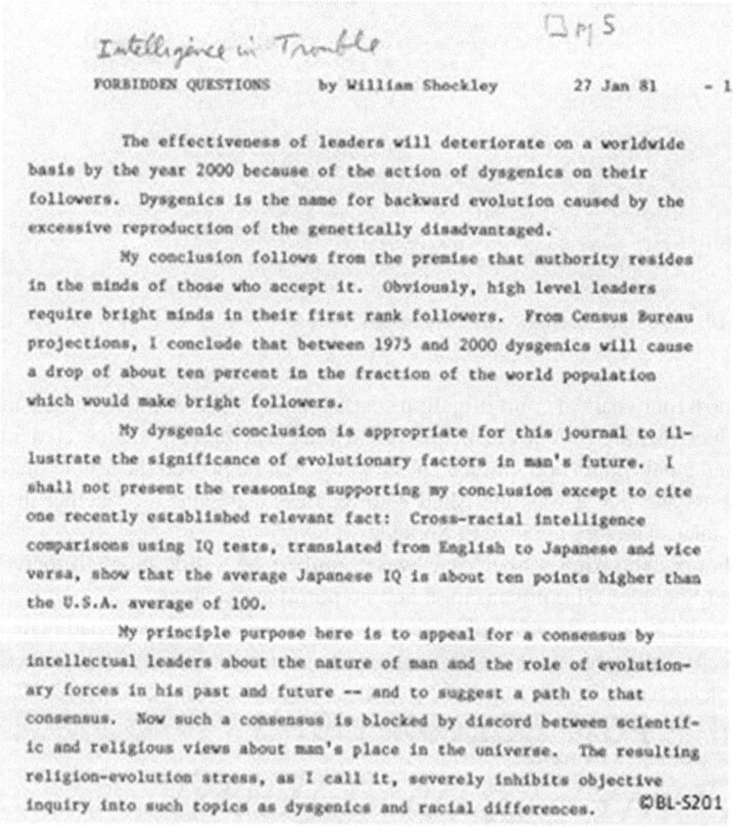

Intelligence in Trouble   □ ℳ S

FORBIDDEN QUESTIONS   by William Shockley   27 Jan 81   - 1

The effectiveness of leaders will deteriorate on a worldwide basis by the year 2000 because of the action of dysgenics on their followers. Dysgenics is the name for backward evolution caused by the excessive reproduction of the genetically disadvantaged.

My conclusion follows from the premise that authority resides in the minds of those who accept it. Obviously, high level leaders require bright minds in their first rank followers. From Census Bureau projections, I conclude that between 1975 and 2000 dysgenics will cause a drop of about ten percent in the fraction of the world population which would make bright followers.

My dysgenic conclusion is appropriate for this journal to illustrate the significance of evolutionary factors in man's future. I shall not present the reasoning supporting my conclusion except to cite one recently established relevant fact: Cross-racial intelligence comparisons using IQ tests, translated from English to Japanese and vice versa, show that the average Japanese IQ is about ten points higher than the U.S.A. average of 100.

My principle purpose here is to appeal for a consensus by intellectual leaders about the nature of man and the role of evolutionary forces in his past and future -- and to suggest a path to that consensus. Now such a consensus is blocked by discord between scientific and religious views about man's place in the universe. The resulting religion-evolution stress, as I call it, severely inhibits objective inquiry into such topics as dysgenics and racial differences.   ©BL-S201

Fig. 14.20 W. Shockley, page 1 of the unpublished article "*Forbidden questions*", January 27, 1961

concerned about human behavior and strived to understand it. A shocking decline in public behavior and discourse led him to the warning that in well-established democracies, like the United States, negative demographic trends would result in inexorable decline, and that democracy would eventually fail. Shockley's manuscript for the article "Forbidden Questions" (Fig. 14.20) summarizing his thoughts was rejected for publication by four journal editorial boards and was never published. However, he never abandoned the topic of dysgenics, pursuing every opportunity to expound his views on it for the rest of his life. Because of his obsession with this "sensitive" view on genetics, which is that heredity rather than environment is the primary cause of the relative disadvantage of certain populations in society, many of his friends and colleagues increasingly avoided contact with him as he aged.

Shockley's view has gradually been confirmed by recent studies of the biological bases of intelligence.[22] Anyone imposed to this idea should ask why, of the billions of people that have lived on planet Earth since the dawn of civilization, the number of truly significant individuals is only a few thousand. But most importantly, social scientists should explain who is responsible for the environment in which each person lives. In the early 20th century, the English biostatistician Karl Pearson discerned on the basis of an impressive array of calculations with high degrees of significance that "*the influence of environment was not one-fifth that of heredity*" and that "*it is man who makes the environment, and not the environment which makes the man.*"

This chapter should provide the reader with more than enough to conclude that his "last five minutes," which he mentioned in his presentation at the NAS were not just as he had hoped they would be. So, I shall let the reader answer the question of Shockley's prophecies for himself.

---

[22]R. J. Haier, "*Biological bases of intelligence*", in Cambridge Handbook of Intelligence, Cambridge University Press, 2011.

Chapter 15
Epilogue

"Every positive value has its price in negative terms, and you never see anything very great which is not, at the same, time, horrible in some respect. The genius of Einstein leads to Hiroshima"

Pablo Picasso to Francoise Gilot, 1948

In 1987 Shockley learned that he had prostate cancer but choose not to undergo surgery. Within a year, the cancer had metastasized to his bones and he began receiving treatments at the Mayo Clinic, although to little avail. He died at home on the morning of August 12, 1989, aged 79.

Shockley, as any mortal person, exhibited a mixed assemblage of both positive and negative character qualities. However, few, if any, people have had a greater impact on the way we are living today than William B. Shockley. Yet, because he did not take things for granted and touched upon subjects which the U.S. National Labor Relations Board in 2018 had characterized as *"harmful, discriminatory, and disruptive,"*[1] Shockley's contributions to mankind have been marginalized and even outright vilified. Fortunately, science does not care what politicians believe. In one of his presentations,[2] Shockley once said: *"In science, all workers are on an equal footing and their accomplishments are judged finally in the impartial court of nature which always operates by the same laws no matter whose experiments they govern."*

Shockley, for his unique and intellectually intense personality, quickly earned a reputation for scientific brilliance. He could look at a problem and already have solved it, while others would still be struggling even to understand where the problem was. It did not matter in which field Shockley worked, he always excelled. When something interested him, that interest was all-consuming and anyone involved in discussing the subject with him would instantly be put on the spot. Such discussions could be grueling and difficult, but always highly instructive. Those who knew Shockley soon learned that he was incapable of insincerity and always made it clear where he stood. He could at times be insulting and ruthless but, for those who accepted

---

[1] J. Damore, D. Gudeman vs. Google, 18CV321529, January 8, 2018.

[2] W. Shockley, *"Crystals, Electronics and Man's Conquest of Nature"*, Int. Conference on Solid-State Physics in Electronics and Telecommunications, Brussels, June 2–7, 1958.

© The Author(s), under exclusive license to Springer Nature Switzerland AG 2021
B. Lojek, *William Shockley: The Will to Think*, Springer Biographies,
https://doi.org/10.1007/978-3-030-65958-5_15

Fig. 15.1 William B. Shockley one year before his death

his personality, his accurate criticism would help them to understand the issue more fully.

Shockley differed from most of those who have written and talked about him in that he always had something important to say.

Shockley's view was that public welfare, rather than alleviating the cause of poverty, might actually be increasing the chain of enslavement of the poor. But politically correct politicians and their followers[3] are only willing to tolerate any dissent as long as it not "wrong," that is, as long as it does not disagree with or question some current dogma.

So, who was this William Shockley, who could recite for hours his beloved poet T.S. Elliot? He was an eccentric, whatever that means. A nonconformist, a difficult man to deal with. His guesses were right so often that it could be unnerving.

It is tempting to use modern psychological labels to describe Shockley, but he was infinitely more complex than any simple label. For Shockley, work became an obsession. Moreover, beyond his passion for science, his early life was marked by

---

[3]To be precise, a more appropriate term would be German term "*Mitläufer*."

Fig. 15.2 Amateur magician W. Shockley at Jim Fisk's retirement party (1973)

his extraordinarily intellectual parents. He was a hard worker who imposed strict self-discipline, a way of life which produced both brilliant insights and sometimes odd behavior. He could be extremely charming and then become totally withdrawn. He was often criticized for being harsh, arrogant, and egoistical. He had a hard time working with others, especially those he perceived as having limited mental abilities. Yet, he also had a magnetic personality, a kind of distinguished sweetness, sincerity, modesty, refinement, generosity, and force, all of which could be quite mesmerizing, and he was more than capable of conveying his enthusiasm for discovery.

When asked about his weaknesses, Shockley said[4]: *"Sometimes I become angry too easily, and this bothers me. It takes me too long to get over it, and this is a waste of time."* His strongest point was his passion for problem solving. He did not give up easily. He did not divide time between his work and recreation. His work was his recreation and his satisfaction. His sports activities always resembled work. When he was interested in mountain climbing, he did not regard it as relaxation; before he embarked on a mountain, he would train rigorously at home. When Shockley courted Emmy, he rented a boat with one sail in Woods Hole, Massachusetts, and invited her to go sailing with him. Shockley arrived with a map, compass, and protractor, while Emmy brought a packed lunch. When they were out on the ocean, Emmy complimented Bill on how well he was handling the boat. Bill's reply was a little slow: *"well I study a lot of books."* It was actually the first time he was out on the open ocean.

The other unique characteristic of Shockley was his good sense of humor. Being an accomplished amateur magician, in a talk at the American Physical Society, Shockley paid a compliment to Dr. Carl Zener, then the Director of Research at

---

[4]S. Thomas, Men of Space, Chilton, Philadelphia, 1960.

Westinghouse. Zener, in his work, had predicted the effect which Shockley studied at Bell Laboratories. This phenomenon involved a particular mechanism which is now called the Zener effect. At an appropriate time, Shockley pulled a bouquet of roses from his sleeve and presented it to Dr. Zener.

As with many other truly exceptional scientists, he frequently came up with unusual ideas. For example, he suggested that all politicians should take a lie detector test. Shockley, inspired by Herbert Hoover's view that: *"The great human advances have not been brought by mediocre men and women. There exists in this country today, a cult of mediocrity which caters to the prejudice that no one person can be much more able than another"*, argued that the cult of mediocrity arises in part from lack of information. Well before the advent of social media, Shockley warned that: *"Cult of mediocrity also arises from what would be called 'noise' in an electronic machine. 'Noise' is due to uncontrolled, false signals which are confused with the true error signals."* He proposed the establishment of a new government department: "Institute of Public Enlightenment" with the mission of finding the most important facts to put before people. Shockley says[5]: *"I advocate the use of the skills and media of advertising in an extensive program of public education on important national issues, such as science teaching in our high schools and the level of teachers' salaries throughout our whole educational structure; the meaning of and need for basic research in the sciences, in human relations, and in mental health; the problems of civil service."* Such an enlightened stance on these various social issues would be considered unreservedly progressive by today's standards.

When Shockley turned his attention to the relationship between heredity and intelligence, his analysis of government data convinced himself that negative dysgenic changes had already occurred in American society. This was when Shockley began his crusade and asked the National Academy of Sciences to address the problem He advocated the establishment of an impartial, nationally funded scientific and not ideological, program of research to identify the root causes of the social problems of the time. He was not asking whether the heritability of intelligence was the same or different in the black or white populations. He presented the evidence that heritability has the same trend in all racial populations. Shockley's analysis led him to this conclusion[6]: *"The less intelligent who inhabited America's miserable inner-city slums were reproducing at above-replacement levels. Inevitably, with the resultant increase in the proportion of people of low intelligence in the population, and the corresponding decline in the numbers with high intelligence, prosperity would be harder to maintain and poverty would surely become more widespread in future generations."*

---

[5]W. Shockley, *"Physics, Dollars, And Sputnik"*, presentation at American Physical Society, December 20, 1957.

[6]W. Shockley, *"Population Control and Eugenics"*, 1956 Nobel Conf. on Genetics and the Future of Man, St. Peter, Minnesota, 1956.

The human suffering that he witnessed during his war experience led him to a second strong concern about the world's population explosion. How was it possible that the prevention of human suffering could become associated with evil?

Shockley wrote in Presbyterian Life on February 1, 1972: *"I came gradually— it took several years—to the moral postulates that lead to my harsh appraisal of our nation's intellectual climate. To start with, I simply operated on the instinctive basis that truth was* per se *a good thing and should be sought no matter what. This position brought me promptly into conflict with what should be but is not our nation's scientific conscience."*

Shockley's effort to address the issues that concerned him could not have had worse timing. He was viciously attacked by Marxist leftists and liberals in the same way as by the God of conservatives at that time, William F. Buckley. It is human nature to jump to conclusions even before the evidence is in, and as no one studied Shockley's data and his analyses, because they could not dispute the data, they attacked him personally. Since in a politically correct society, a high professional reputation depends upon a sustained history of "appropriate" behavior, American academia preferred to live with a lie. While no scientist ever proved that Shockley was wrong, neither did any join him in support. The truth was not that all scientists were political cowards; they just did not know how to be political heroes.

The previous chapters in this book may have left the reader wondering how much or how little of history they can trust. As any voracious reader of history will know, historians have never been able to reconcile two eyewitness accounts of any given event.

The FBI Report characterizes Shockley in many entries as[7] *"'good all the way' he has a conscious concept of giving unstintingly his time for the benefit of his country. He is a person of highest standards who expected the same of his associates."* So, I asked a few scientists who worked with Shockley for help. My question was how they would characterize William Shockley in a few words. Here are the results:

Jack A. Morton: *"During the course of my own career, I have had the opportunity to study under, or work with, quite a number of Nobel Prize winners in physics. I think I can say without equivocation that of them all Shockley stands out in combining outstanding strength in teaching, in leadership, and in independent contributions."*

John Pierce said: *"I owe a great deal to him, for I learned a lot of what I know about physics at the Bell Laboratories, just through his kindness in helping me after I went there. Shockley really loves science and takes all of his interests very seriously."*

Morgan Sparks, who worked with all three transistor inventors, provided a brisk answer: *"Bardeen and Brattain were conventional physicists, Shockley was unconventional. He has the ability to recognize and ignore the many aspects of the problem that are not of great importance and to analyze the relevant factors quickly and in the simplest possible way. I believe he fills all the qualifications to be considered a true genius."*

Phil Anderson provided a longer answer: *"Bell Laboratories was a unique research place with a highly competitive environment. Contrary to common beliefs,*

---

[7]Page SF161-479 in Shockley's FBI Report.

Bardeen had a temper. Abrasive and often cruel, Shockley could be a charming man if you learn how to challenge him."

Ian Ross wrote in an e-mail: *"Bell laboratories did not have another physicist of Shockley's caliber. Looking back on Shockley's career, I believe that the most unfortunate event was his decision to leave Bell Labs. Bill deserved a higher salary and a more exalted position."*

Kurt Hübner wrote in a letter to the author: *"Came a period, when Shockley spent hours on hours trying to explain basic physics to lesser educated employees in order to get an idea on how their thinking worked. He always had an immense patience with people who admitted not to understand, while people pretending to understand, but didn't, found no sympathy."*

Esther Conwell added: *"Shockley was always nice to me. He was a gentleman."*

Harry Sello stated: *"Shockley was a brilliant genius. I have never been more stimulated technically than in three years working with Shockley."*

The above characterizations contrast sharply with descriptions of William Shockley after the desertion of the "California Group" in 1957, which could be found in current literature or posts available on the Internet. There are at least two reasons for such discrepancies: one is offered by Hermann Göring in the quote at the beginning of Chap. 11. Such an explanation would apply to historians. The second reason is that portraying Shockley as a racist allowed some to distract from the questionable behavior of people who are now portrayed as heroes.

As I mentioned earlier, I am not an historian. I spent my entire professional career in the semiconductor business and I am convinced that Shockely's reputation was destroyed because Shockley asked questions that no one wanted to ask, much less answer. Shockley trespassed on a taboo subject. After taping David Frost's TV show in London, Emmy told Frost: *"He is not so much angry as frustrated that he has been misunderstood in his view on genetics."*

I think we all misunderstood Shockley. We thought he was a dreamer and visionary. He did dream and his visionary thoughts came true; he had visions but they were of a real future. He is revered for offering disruptive insights, particularly when today's technological advances tend to be incremental. Some of his principle beliefs were that technology should transcend the marketplace and that innovation should not just be tied to profit. Driven by inner forces, he never accepted the status quo.

Shockley, with his explosive thought process, was unable to convince the public that it would be advisable to carry out a careful, thoughtful, and objective study of genetic effects on intelligence. If he was wrong, then it would be well to lay his conjectures aside for all time. If he was right, then it is better that we know. We can argue about whether his "style" was the best way to get the message out. William Shockley may have been wrong, but all his theories were in the end supported by the weighty results of science. The ultimate question the reader should ask is: *"What if he was right?"* An impartial court might appreciate William B. Shockley better in a few hundred years from now, when all his detractors will have sunk into oblivion and when we will have learnt from the mistakes and damage that political correctness can sometimes inflict upon us.

Chapter 16
Appendix: Shockley's Patents, Papers and Presentations

16.1 Patents

1. W. Shockley, "Electron discharge device," U.S. Patent 2,207,355, July 9, 1940.
2. W. Shockley, "Electron discharge device," U.S. Patent 2,236,012, Mar. 25, 1941.
3. J. R. Pierce and W. Shockley, "Electron multiplier," U.S. Patent 2,245,605, June 17, 1941.
4. W. Shockley, "Electron discharge device," U.S. Patent 2,245,614, June 17, 1941.
5. W. Shockley and G. W. Willard, "Wave propagation device," U.S. Patent 2,407,294, Sept. 10, 1946.
6. W. Shockley, "Ultra high frequency electronic device," U.S. Patent 2,409,227, Oct. 15, 1946.
7. W. Shockley, "Pulse generator," U.S. Patent 2,416,718, Mar. 4, 1947.
8. G. L. Pearson and W. Shockley, "Semiconductor amplifier," U.S. Patent 2,502,479, Apr. 4, 1950.
9. W. Shockley, "Semiconductor amplifier," U.S. Patent 2,502,488, Apr. 4, 1950.
10. W. Shockley, "Differential altimeter," U.S. Patent 2,509,889, May 30, 1950.
11. W. Shockley, "Acoustic transducer utilizing semiconductors," U.S. Patent 2,553,491, May 15, 1951.
12. W. Shockley, "Circuit element utilizing semiconductive materials," U.S. Patent 2,569,347, Sept. 25, 1951.
13. J. R. Haynes and W. Shockley, "Semiconductor signal translating device with controlled carrier transit times," U.S. Patent 2,600,500, June 17, 1952.
14. W. Shockley, "Circuit element utilizing semiconductive materials," U.S. Patent 2,623,102, Dec. 23, 1952.
15. W. Shockley and M. Sparks, "Semiconductor translating device having controlled gain," U.S. Patent 2,623,105, Dec. 23, 1952.
16. W. Shockley, "Semiconductor signal translating device," U.S. Patent 2,654,059, Sept. 29, 1953.

© The Author(s), under exclusive license to Springer Nature Switzerland AG 2021
B. Lojek, *William Shockley: The Will to Think*, Springer Biographies,
https://doi.org/10.1007/978-3-030-65958-5_16

17. W. Shockley, "Bistable circuits including transistors," U.S. Patent 2,655,609, Oct. 13, 1953.
18. W. Shockley, "Semiconductor translating device," U.S. Patent 2,666,814, Jan. 19, 1954.
19. W. Shockley, "Transistor amplifier," U.S. Patent 2,666,818, Jan. 19, 1954.
20. W. Shockley, "Semiconductor translating device," U.S. Patent 2,672,528, Mar. 16, 1954.
21. W. Shockley, "Circuit element utilizing semiconductive materials," U.S. Patent 2,681,993, June 22, 1954.
22. H. B. Briggs, J. R. Haynes, and W. Shockley, "Infrared energy source," U.S. Patent 2,683,794, July 18, 1954.
23. W. Shockley, "Electrooptical control system," U.S. Patent 2,696,565, Dec. 7, 1954.
24. W. Shockley, "Circuits, including semiconductor device," U.S. Patent 2,714,702, Aug. 2, 1955.
25. W. Shockley, "Transistor circuits with constant output current," U.S. Patent 2,716,729, Aug. 30, 1955.
26. W. Shockley, "Method of making semiconductor crystals," U.S. Patent 2,730,470, Jan. 10, 1956.
27. W. Shockley, "Semiconductor signal translating devices," U.S. Patent 2,744,970, May 8, 1956.
28. W. Shockley, "Semiconductor signal translating devices," U.S. Patent 2,756,285, July 24, 1956.
29. W. Shockley, "Frequency selective semiconductor circuit elements," U.S. Patent 2,761,020, Aug. 28, 1956.
30. W. Shockley, "Semiconductor circuit controlling device," U.S. Patent 2,763,832, Sept. 18, 1956.
31. W. Shockley, "Semiconductor signal translating devices," U.S. Patent 2,764,642, Sept. 25, 1956.
32. W. Shockley, "Negative resistance device application," U.S. Patent 2,772,360, Nov. 27, 1956
33. W. P. Mason and W. Shockley, "Negative resistance amplifiers," U.S. Patent 2,775,658, Dec. 25, 1956
34. W. Shockley, "Asymmetric waveguide structure," U.S. Patent 2,777,906, Jan. 15, 1957.
35. W. Shockley, "Semiconductor signal translating devices," U.S. Patent 2,778,885, Jan. 22, 1957.
36. W. Shockley, "Forming semiconductive devices by ionic bombardment," U.S. Patent 2,787,564, Apr. 2, 1957.
37. W. Shockley, "Semiconductor signal translating devices," U.S. Patent 2,790,037, Apr. 23, 1957.
38. W. Shockley, "Nonreciprocal circuits employing negative resistance elements," U.S. Patent 2,794,864, June 4, 1957.
39. W. Shockley, "High frequency negative resistance device," U.S. Patent 2,794,917, June 4, 1957.

40. W. Shockley, "Semiconductive device," U.S. Patent 2,813,233, Nov.12, 1957.
41. W. Shockley, "Method of fabricating semiconductor signal translating devices," U.S. Patent 2,816,847, Dec. 17, 1957.
42. W. Shockley, "High frequency negative resistance device," U.S. Patent 2,852,677, Sept. 16, 1958.
43. W. Shockley, "Semiconductive switch," U.S. Patent 2, 855,524, Oct. 7, 1958.
44. W. Shockley, "Method of forming large area P-N junctions," U.S. Patent 2, 868,678, Jan. 13, 1959.
45. W. Shockley, "Negative resistance semiconductive device," U.S. Patent 2,869,084, Jan. 13, 1959.
46. W. Shockley, "Method for growing junction semiconductive devices," U.S. Patent 2,879,189, Mar. 24, 1959.
47. W. Shockley, "Radiant energy control system," U.S. Patent 2,884,540, Apr. 28, 1959.
48. W. Shockley, "Semiconductive material purification method and apparatus," U.S. Patent 2,890,139, June 9, 1959.
49. W. Shockley, "Transistor switch," U.S. Patent 2,891,171, June 16, 1959.
50. W. Shockley, "Shifting register," U.S. Patent 2,912,598, Nov. 10, 1959.
51. W. Shockley, "Crystal growing apparatus," U.S. Patent 2,927,008, Mar.1, 1960.
52. W. Shockley, "Semiconductor amplifying device," U.S. Patent 2 936425, May 10, 1960.
53. W. Shockley, "Semiconductive device and method," U.S. Patent 2,937,114, May 17, 1960.
54. W. Shockley, "P-N junction having minimum transition layer capacitance," U.S. Patent 2,953,488, Sept. 20, 1960.
55. W. Shockley, "Grain boundary semiconductor device and method," U.S. Patent 2,954,307, Sept. 27, 1960.
56. W. Shockley, "Semiconductor shift register," U.S. Patent 2,967,952, Jan. 10, 1961.
57. W. Shockley and R. N. Noyce, "Transistor structure," U.S. Patent 2,967,985, Jan. 10, 1961.
58. W. Shockley and R. V. Jones, "Crystal growing apparatus," U.S. Patent 2,979,386, Apr. 11, 1961.
59. W. Shockley, "Semiconductor device and method of making the same," U.S. Patent 2,979,427, Apr. 11, 1961.
60. W. Shockley, "Junction transistor," U.S. Patent 2,980,830, Apr. 18, 1961.
61. W. Shockley, "Fabrication of semiconductor elements," U.S. Patent 2,982,002, May 2, 1961.
62. W. Shockley, "Method of making a semiconductive switching array," U.S. Patent 2,994,121, Aug. 1, 1961.
63. W. Shockley "Semiconductive device and method of operating same," U.S. Patent 2,997,604, Aug. 22, 1961.
64. W. Shockley, "Semiconductor devices," U.S. Patent 3,015,762, Jan. 2, 1962.
65. G. C. Dacey, C. A. Lee, and W. Shockley, "Semiconductive device and method of operating same," U.S. Patent 3,028, 655, Apr. 10, 1962.

66. W. Shockley, "Process for growing single crystals," U.S. Patent 3,031,275, Apr. 24, 1962.

67. W. Shockley, "Semiconductive wafer and method of making the same," U.S. Patent 3,044,909, July 17, 1962.

68. W. Shockley, "Reverse breakdown diode pulse generator," U.S. Patent 3,048,710, Aug. 7, 1962.

69. W. Shockley, "Method of growing silicon carbide crystals," U.S. Patent 3,053,635, Sept. 11, 1962.

70. W. Shockley "Trigger circuit switching from stable operation in the negative resistance region to unstable operation," U.S. Patent 3,058,009, Oct. 9, 1962.

71. W. Shockley, "Electrical component holder," U.S. Patent 3,076,170, Jan. 29, 1963.

72. W. Shockley and A. Goetzberger, "Thermostat," U.S. Patent 3,079,484, Feb. 26, 1963.

73. W. Shockley and A. O. Beckman, "Semiconductor leads and method of attaching," U.S. Patent 3,086,281, Apr. 23, 1963.

74. W. Shockley, "Method of making thin slices of semiconductive material," U.S. Patent 3,096,262, July 2, 1963.

75. W. Shockley, "Semiconductive device," U.S. Patent 3,099,591, July 30, 1963.

76. W. Shockley, "Field effect transistor having grain boundary therein," U.S. Patent 3,126,505, Mar. 24, 1964.

77. W. Shockley and R. N. Noyce, "Method of making a transistor structure," U.S. Patent 3,140,206, July 7, 1964.

78. W. Shockley and G. S. Horsley, "Voltage regulating semiconductor device," U.S. Patent 3,140,438, July 7, 1964.

79. W. Shockley, "Voltage regulating semiconductor device," U.S. Patent 3,154,692, Oct. 27, 1964.

80. G. C. Dacey, C. A. Lee, and W. Shockley, "Mesa transistor with impurity concentration in the base decreasing toward collector junction," U.S. Patent 3,202,887, Aug. 24, 1965.

81. W. Shockley, "Semiconductive device and method of making the same," U.S. Patent 3,236,698, Feb. 22, 1966.

82. W. Shockley, "Thermally stabilized semiconductor device," U.S. Patent 3,286,138, Nov. 15, 1966.

83. W. Shockley, "Surface controlled avalanche transistor," U.S. Patent 3,339,086, Aug. 29, 1967.

84. W. Shockley and R. H. Haitz, "Noise diodes," U.S. Patent 3,349,298, Oct. 24, 1967.

85. W. Shockley and D. R. Curran, "Piezoelectric resonator," U.S. Patent 3,384,768, May 21, 1968.

86. W. Shockley, "Semiconductor device having regions of different conductivity types wherein current is carried by the same type of carrier in all said regions," U.S. Patent 3,398,334, Aug. 20, 1968.

87. A. H. Bobeck, U. F. Gianola, R. C. Sherwood, and W. Shockley, "Magnetic domain propagation circuit," U.S. Patent 3,460,116, Aug. 5, 1969.

88. A. H. Bobeck, P. C. Michaelis, and W. Shockley, "Readout implementation for magnetic memory," U.S. Patent 3,508, 222, Apr. 21, 1970.

89. A. H. Bobeck, H. E. D. Scovil, and W. Shockley, "Magnetic logic arrangement," U.S. Patent 3,541,522, Nov. 17, 1970.

90. J. A. Davis and W. Shockley, "Chopper devices and circuits," U.S. Patent 3,808,515, Apr. 30, 1974.

16.2 Papers

1.   R. P. Johnson and W. Shockley, "An electron microscope for filaments: Emission and absorption by tungsten single crystals," Phys. Rev., vol. 49, pp. 436–440, Mar. 15, 1936.

2.   W. Shockley, "Application of an electrical timing device to certain mechanics experiments," Amer. Phys. Teacher, vol. 4, pp. 76–81, May 1936.

3.   J. C. Slater and W. Shockley, "Optical absorption by the alkali halides," Phys. Rev., vol. 50, pp. 705–719, Oct. 15, 1936.

4.   W. Shockley, "Electronic energy bands in sodium chloride," Phys. Rev., vol. 50, pp. 754–759, Oct. 15, 1936.

5.   J. B. Fisk, L. I. Schiff, and W. Shockley, "On binding of neutrons and protons," Phys. Rev., vol. 50, no. 11, pp. 1090–1091, Dec. 1, 1936.

6.   W. Shockley, "Energy bands for the face-centered lattice," Phys. Rev., vol. 51, pp. 129–135, Jan. 15, 1937.

7.   W. Shockley, "The empty lattice test of the cellular methods in solids," Phys. Rev., vol. 52, no. 8, pp. 866–872, Oct. 15, 1937.

8.   W. Shockley, *Theory of order for the copper gold alloy system,* J. Chem. Phys., vol. 6, no. 3, pp. 130–144, Mar. 1938.

9.   F. C. Nix and W. Shockley, *Order-disorder transformations in alloys,* Rev. Modern Phys., vol. 10, no. 1, pp. 1–71, Jan. 1938.

10. W. Shockley and J. R. Pierce, *A theory of noise for electron multipliers,* Proc. IRE, vol. 26, pp. 321–332, Mar. 1938.

11. C. E. Fay, A. L. Samuel, and W. Shockley, *On the theory of space charge between parallel plane electrodes,* Bell Syst. Tech. J., vol. XVII, pp. 49–79, Jan. 1938.

12. W. Shockley, *On the interaction of atoms in alloys,* J. Chem. Phys., vol. 6, no. 9, pp. 523–525, Sept. 1938.

13. W. Shockley, *Currents to Conductors induced by a moving point charge,* J. Appl. Phys., vol. 9, no. 10, pp. 635–636, Oct. 1938.

14. J. Steigman, W. Shockley, and F. C. Nix, *The self-diffusion of copper,* Phys. Rev., 2nd series, vol. 56, pp. 13–21, July 1, 1939.

15. W. Shockley, *Nature of the metallic state,* J. Appl. Phys., vol. 10, no. 8, pp. 543–555, Aug. 1939.

16. W. Shockley, *On the surface states associated with a periodic potential,* Phys. Rev., vol. 56, no. 4, pp. 317–323, Aug. 15, 1939.

17.   W. Shockley, *"The quantum physics of solids, I—The energies of electrons in crystals,"* Bell Syst. Tech. J., vol. XVIII, pp. 645–723, Oct. 1939.

18.   J. Bardeen, W. H. Brattain, and W. Shockley, *"Investigation of oxidation of copper by use of radioactive Cu tracer,"* J. Chem. Phys., vol. 14, no. 12, pp. 714–721, Dec. 1946.

19.   W. H. Brattain and W. Shockley, *"Density of surface states on silicon deduced from contact potential measurements,"* Phys. Rev., vol. 72, no. 4, p. 345, Aug. 15, 1947.

20.   R. D. Heidenreich and W. Shockley, *"Electron microscope and electron-diffraction study of slip in metal crystals,"* J. Appl. Phys., vol. 18, no. 11, pp. 1029–1031, Nov. 1947.

21.   W. Shockley, *"Study of slip in aluminum crystals by electron microscope and electron diffraction methods,"* in Proc. Conf. Strength of Solids, H. H. Wills Physical Laboratory, Univ. Bristol, U.K., July 7–9, 1947, pp. 57–75. Reprinted in Bell Telephone System Tech. Pub., Monograph B-1618, pp. 1–21.

22.   J. R. Haynes and W. Shockley, *"The trapping of electrons in silver chloride,"* in Physical Society Bristol Conf. Rep., London, U.K., 1948, pp. 151–157.

23.   W. P. Mason, H. J. McSkimin, and W. Shockley, *"Ultrasonic observation of twinning in tin,"* Phys. Rev., vol. 73, no. 10, pp. 1213–1214, May 15, 1948.

24.   W. Shockley and G. L. Pearson, *"Modulation of conductance of thin films of semi-conductors by surface charges,"* Phys. Rev., vol. 74, pp. 232–233, July 15, 1948.

25.   H. J. Williams, R. M. Bozorth, and W. Shockley, *"Magnetic domain patterns on single crystals of silicon iron,"* Phys. Rev., vol. 75, no. 1, pp. 155–178, Jan. 1, 1949.

26.   H. J. Williams and W. Shockley, *"A simple domain structure in an iron crystal showing a direct correlation with the magnetization,"* Phys. Rev., vol. 75, pp. 178–183, Jan. 1, 1949.

27.   E. J. Ryder and W. Shockley, *"Interpretation of dependence of resistivity of germanium on electric field,"* Phys. Rev., vol. 75, p. 310, Jan. 15, 1949.

28.   J. R. Haynes and W. Shockley, *"Investigation of hole injection in transistor action,"* Phys. Rev., vol. 75, p. 691, Feb. 15, 1949.

29.   W. Shockley and W. T. Read, *"Quantitative predictions from dislocation models of crystal grain boundaries,"* Phys. Rev., vol. 75, no. 4, p. 692, Feb. 15, 1949.

30.   H. Suhl and W. Shockley, *"Concentrating holes and electrons by magnetic fields,"* Phys. Rev., vol. 75, no. 10, pp. 1617–1618, May 15, 1949.

31.   W. Shockley, *"Dislocation theory, in cold working of metals,"* Amer. Soc. Metals, pp. 131–147, 1949.

32.   W. Shockley and J. Bardeen, *"Energy bands and mobilities in monatomic semiconductors,"* Phys. Rev., vol. 77, no. 3, pp. 407–408, Feb. 1, 1950.

33.   W. Shockley, G. L. Pearson, and J. R. Haynes, *"Hole injection in germanium - Quantitative studies and filamentary transistors,"* Bell Syst. Tech. J., vol. XXVIII, no. 3, pp. 344–366, July 1949.

34. W. Shockley, *"The theory of p-n junctions in semiconductors and p-n junction transistors,"* Bell Syst. Tech. J., vol. XXVIII, no. 3, pp. 435–489, July 1949.

35. C. Kittel, E. A. Nesbitt, and W. Shockley, *"Theory of magnetic properties and nucleation in alnico V,"* Phys. Rev., vol. 77, no. 6, pp. 839–840, Mar. 15, 1950.

36. W. Shockley, *"Energy band structures in semiconductors,"* Phys. Rev., vol. 78, no. 2, pp. 173–174, Apr. 15, 1950.

37. W. T. Read and W. Shockley, *"Dislocation models of crystal grain boundaries,"* Phys. Rev., vol. 78, no. 3, pp. 275–289, May 1, 1950.

38. W. Shockley, *"Theories of high values of alpha for collector contacts on germanium,"* Phys. Rev., vol. 78, no. 3, pp. 294–295, May 1, 1950.

39. G. L. Pearson, J. R. Haynes, and W. Shockley, *"Comment on mobility anomalies in germanium,"* Phys. Rev., vol. 78, no. 3, pp. 295–296, May 1, 1950.

40. W. Shockley, *"Effect of magnetic fields on conduction—Tube integrals,"* Phys. Rev., vol. 79, no. 1, pp. 191–192, July 1, 1950.

41. W. Shockley, *"Holes and electrons,"* Phys. Today, vol. 3, no. 10, pp. 16–24, Oct. 1950.

42. J. Bardeen and W. Shockley, *"Scattering of electrons in crystals in the presence of large electric fields,"* Phys. Rev., vol. 80, no. 1, pp. 69–71, Oct. 1, 1950.

43. W. Shockley, *"Deformation potentials and mobilities in non-polar crystals,"* Phys. Rev., vol. 80, no. 1, pp. 72–80, Oct. 1, 1950.

44. H. J. Williams, W. Shockley, and C. Kittel, *"Studies of the propagation velocity of a ferromagnetic domain boundary,"* Phys. Rev., vol. 80, no. 6, pp. 1090–1094, Dec. 15, 1950.

45. E. J. Ryder and W. Shockley, *"Mobilities of electrons in high electric fields,"* Phys. Rev., vol. 81, no. 1, pp. 139–140, Jan. 1, 1951.

46. F. S. Goucher, G. L. Pearson, M. Sparks, G. K. Teal, and W. Shockley, *"Theory and experiment for a germanium p-n junction,"* Phys. Rev., vol. 81, no. 4, pp. 637–638, Feb. 15, 1951.

47. J. R. Haynes and W. Shockley, *"The mobility and life of injected holes and electrons in germanium,"* Phys. Rev., vol. 81, no. 5, pp. 835–843, Mar. 1, 1951.

48. J. R. Haynes and W. Shockley, *"The mobility of electrons in silver chloride,"* Phys. Rev., vol. 82, no. 6, pp. 935–943, June 15, 1951.

49. W. Shockley, M. Sparks, and G. K. Teal, *"p-n junction transistors,"* Phys. Rev., vol. 83, no. 1, pp. 151–162, July 1, 1951.

50. W. Shockley, *"Hot electrons in germanium and Ohm's law,"* Bell Syst. Tech. J., vol. XXX, no. 4, part 1, pp. 990–1034, Oct. 1951.

51. K. B. McAfee, E. J. Ryder, W. Shockley, and M. Sparks, *"Observations of Zener current in germanium p-n junctions,"* Phys. Rev., vol. 83, no. 3, pp. 650–651, Aug. 1, 1951.

52. W. Shockley, *"New phenomena of electronic conduction in semi-conductors,"* Phys. Today, pp. 26–36, 1951.

53. G. L. Pearson, W. T. Read, and W. Shockley, *"Probing the space-charge layer in a p-n junction,"* Phys. Rev., vol. 85, no. 6, pp. 1055–1057, 15th Mar. 1952.

54. K. B. McAfee, W. Shockley, and M. Sparks, *"Measurement of diffusion in semiconductors by a capacitance method,"* Phys. Rev., vol. 86, no. 1, pp. 137–138, Apr. 1, 1952.

55. W. T. Read, Jr., and W. Shockley, *"On the geometry of dislocations,"* in Imperfections in Nearly Perfect Crystals. New York: Wiley, 1952, Ch. 2, pp. 77–94.

56. W. Shockley, *"Dislocation models of grain boundaries,"* in Imperfections in Nearly Perfect Crystals. New York: Wiley, 1952, Ch. 13, pp. 352–376.

57. W. Shockley, *"Solid state physics in electronics and in metallurgy,"* Trans. AIME, J. Metals, vol. 194, pp. 829–842, Aug. 1952.

58. W. Shockley and W. T. Read, Jr., *"Statistics of the recombination of holes and electrons,"* Phys. Rev., vol. 87, no. 5, pp. 935–842, Sept. 1, 1952.

59. W. Shockley, *"Interpretation of efm values for electrons in crystals,"* Phys. Rev., vol. 88, no. 4, p. 953, Nov. 15, 1952.

60. W. Shockley, *"Transistor electronics: Imperfections, unipolar and analog transistors,* Proc. IRE, vol. 40, pp. 1289–1313, Nov. 1952.

61. W. Shockley, *"A unipolar 'field-effect' transistor,"* Proc. IRE, vol. 40, pp. 1365–1376, Nov. 1952.

62. T. S. Benedict and W. Shockley, *"Microwave observation of the collision frequency of electrons in germanium,"* Phys. Rev., vol. 89, no. 5, pp. 1152–1153, Mar. 1, 1953.

63. J. D. Eshelby, W. T. Read, and W. Shockley, *"Anisotropic elasticity with applications to dislocation theory,"* Acta Metallurgica, vol. 1, pp. 251–259, May 1953.

64. W. Shockley, *"Cyclotron resonances, magnetoresistance, and Brillouin Jones in semiconductors,"* Phys. Rev., vol. 90, no. 3, p. 491, May 1, 1953.

65. W. Shockley and R. C. Prim, *"Space-charge limited emission in semiconductors,"* Phys. Rev., vol. 90, no. 5, pp. 753–758, June 1, 1953.

66. W. Shockley, *"Some predicted effects of temperature gradients on diffusion in crystals,"* Phys. Rev., vol. 91, no. 6, pp. 1563–1564, Sept. 15, 1953.

67. J. R. Tesman, A. H. Kahn, and W. Shockley, *"Electronic polarizabilities of ions in crystals,"* Phys. Rev., vol. 92, no. 4, pp. 890–895, Nov. 15, 1953.

68. R. C. Prim and W. Shockley, *"Joining solutions at the pinch-off in 'field-effect' transistor,"* Trans. IRE, Professional Group Electron Devices, vol. PGED-4, pp. 1–14, Dec. 1953.

69. W. Shockley, "Transistor physics," Amer. Scientist, vol. 42, no. 1, pp. 41–72, Jan. 1954.

70. W. Shockley, *"Some predicted effects of temperature gradients on diffusion in crystals,"* Phys. Rev., vol. 93, no. 2, pp. 345–346, Jan. 15, 1954.

71. W. Shockley and W. P. Mason, *"Dissected amplifiers using negative resistance,"* J. Appl. Phys., vol. 25, no. 5, p. 677, May 1954.

72. W. van Roosebroeck and W. Shockley, *"Photon-radiative recombination of electrons and holes in germanium,"* Phys. Rev., vol. 94, no. 6, pp. 1558–1560, June 15, 1954.

73. W. Shockley, *"Negative resistance arising from transit time in semiconductor diodes,"* Bell Syst. Tech. J., vol. XXXIII, no. 4, pp. 799–826, July 1954.

74. W. Shockley, "Les semi-conducteurs," Le Vide, no. 56, pp. 9–26, Mar./Avr. 1955 (French adaptation of three conferences presented in English at the Sorbonne, Oct. 1953).

75. W. Shockley, *"Transistor physics,"* Proc. Inst. Elect. Eng., vol. 103, pt. B, no. 7, pp. 23–41, Jan. 1956.

76. W. Shockley, *"Localized radiation damage as a means of studying vacancies and interstitials,"* in Dislocations and Mechanical Properties of Crystals, J. C. Fisher, W. G. Johnston, R. Thomson, and T. Vreeland, Jr., Eds. New York: Wiley, 1957, pp. 581–586.

77. W. Shockley, *"Transistor technology evokes new physics,"* Les Prix Nobel 1956, Nobel lecture, Stockholm, Sweden, pp. 100–129, Dec. 11, 1956.

78. W. Shockley, *"On the statistics of individual variations of productivity in research laboratories,"* Proc. IRE, vol. 45, pp. 279–290, Mar. 1957.

79. B. F. Miesner and W. Shockley, *"On the statistics of individual variations of productivity in research laboratories,"* Proc. IRE, vol. 45, pp. 1409–1410, Oct. 1957.

80. P. D. Allison, J. A. Stewart, and W. Shockley, *"Productivity differences among scientists: evidence for accumulative advantage,"* Amer. Soc. Rev., vol. 39, pp. 596–606, Aug. 1974.

81. W. Shockley and J. T. Last, *"Statistics of the charge distribution for a localized flaw in a semiconductor,"* Phys. Rev., vol. 107, no. 2, pp. 392–396, July 15, 1957.

82. W. Shockley, *"The statistics of quality losses in civil service laboratories,"* National Academy of Sciences, National Research Council, NAS-ARDC Special Study COM-4-T19, pp. 1–14, Oct. 1957.

83. W. Shockley, *"Unique properties of the four-layer diode,"* Electron. Ind. Tele Tech, Aug. 1957.

84. C.-T. Sah, R. N. Noyce, and W. Shockley, *"Carrier generation and recombination in P-N junctions and P-N junction characteristics,"* Proc. IRE, vol. 45, pp. 1228–1243, Sept. 1957.

85. W. Shockley and J. F. Gibbons, *"Introduction to the four-layer diode,"* Semiconductor Products, vol. 1, no. 1, pp. 9–13, Jan./Feb. 1958.

86. W. Shockley, "Guest editorial," Semiconductor Products, p. 5, Mar./Apr. 1958.

87. C.-T. Sah and W. Shockley, *"Electron-hole recombination statistics in semiconductors through flaws with many charge conditions,"* Phys. Rev., vol. 109, no. 4, pp. 1103–1115, Feb. 15, 1958.

88. W. Shockley, *"Transistor electronics has good future,"* Ind. Laboratories, pp. 52–53, May 1958.

89.   W. Shockley, *"Electron, holes and traps,"* Proc. IRE, vol. 46, pp. 973–990, June 1958.

90.   W. Shockley, *"An invited essay on transistor business,"* Proc. IRE, vol.46, pp. 954–955, June 1958.

91.   W. Shockley, *"Predicted intervalley scattering effects in germanium,"* Phys. Rev., vol. 110, no. 5, pp. 1207–1208, June 1, 1958.

92.   W. Shockley and J. F. Gibbons, *"Study of ultimate high frequency and high-power limits of semiconductor devices,"* U.S. Army Signal Corps Engineering Laboratories, Fort Monmouth, NJ, Final Rep., July 1, 1957–Oct. 15, 1958.

93.   W. Shockley and J. F. Gibbons, *"Theory of transient build-up in avalanche transistor,"* Commun. Electron., vol. 40, p. 993, Jan. 1959. Reprinted from Solid State Physics in Electronics and Telecommunications. London, U.K.: Academic, 1958, pp. 1024–1035.

94.   W. Shockley and J. F. Gibbons, *"Current build-up in semiconductor devices,"* Proc. IRE, vol. 46, pp. 1947–1949, Dec. 1958.

95.   D. J. Hamilton, J. F. Gibbons, and W. Shockley, *"Physical principles of avalanche transistor pulse circuits,"* Proc. IRE, vol. 47, pp. 1102–1108, June 1959.

96.   A. Goetzberger and W. Shockley, *"Localized excess reverse currents in silicon p-n junctions,"* in Structure and Properties of Thin Films. New York: Wiley, 1959, pp. 298–301.

97.   K. Hubner and W. Shockley, *"Analysis of diffusion down dislocations,"* in Structure and Properties of Thin Films. New York: Wiley, 1959, pp. 302–305.

98.   W. Shockley, *"Theory of transmitted phonon drag,"* in Structure and Properties of Thin Films. New York: Wiley, 1959, pp. 306–327.

99.   W. Shockley, *"The four-layer transistor diode: An example of a solid-state circuit or molecular engineering,"* Wave Guide, vol. X, no. 7, Mar. 1959.

100. W. Shockley, *"Transistor-diodes,"* Proc. Inst. Elect. Eng., vol. 106, pt. B, suppl. 15, pp. 270–272, May 21, 1959.

101. W. Shockley, *"Discussion on basic theory-II (minimum capacity junction),"* Proc. Inst. Elect. Eng., vol. 106, pt. B, suppl. 17, May 22, 1959.

102. K. Hubner and W. Shockley, *"Transmitted phonon drag measurements in silicon,"* Phys. Rev. Lett., vol. 4, no. 10, pp. 504–505, May 15, 1960.

103. W. Shockley, *"Transmitted phonon drag measurements in silicon,"* in Proc. Int. Conf. Semiconductor Physics, Prague, Czechoslovakia, 1960, pp. 229–231.

104. W. Shockley and J. L. Moll, *"Solubility of flaws in heavily-doped semiconductors,"* Phys. Rev., vol. 119, no. 5, pp. 1480–1482, Sept. 1, 1960.

105. A. Goetzberger and W. Shockley, *"Metal precipitates in silicon p-n junctions,"* J. Appl. Phys., vol. 31, no. 10, pp. 1821–1824, Oct. 1960.

106. W. Shockley, *"Problems related to p-n junctions in silicon,"* Solid State Electron., vol. 2, no. 1, pp. 35–67, Jan. 1961.

107. K. Hubner and W. Shockley, *"New experiments on interaction of phonons with crystalline defects,"* in Advanced Energy Conversion. London, U.K.: Pergamon, 1961, vol. 1, pp. 93–96.

108. M. A. Melehy and W. Shockley, "*Response of a p-n junction to a linearly decreasing current*," Trans. IRE, Professional Group Electron Devices, vol. PGED-8, no. 2, pp. 135–139, Mar. 1961.

109. W. Shockley and H. J. Queisser, "*Detailed balance limit of efficiency of p-n junction solar cells*," J. Appl. Phys., vol. 32, pp. 510–519, Mar. 1961. Reprinted in Solar Cells, C. E. Backus, Ed. New York: IEEE Press, 1976, pp. 136–145.

110. H. J. Queisser and W. Shockley, "*Some theoretical aspects of the physics of solar cells*," in Progress in Astronautics and Rocketry, vol. 3, Energy Conversion for Space Power. New York: Academic, 1961, pp. 317–323.

111. H. J. Queisser, K. Hubner, and W. Shockley, "*Diffusion along small-angle grain boundaries in silicon*," Phys. Rev., vol. 123, no. 4, pp. 1245–1254, Aug. 15, 1951.

112. W. Shockley, "*Field enhanced donor diffusion in degenerate semiconductor layers*," J. Appl. Phys., vol. 32, p. 1402, July 1961.

113. W. Shockley, "*Diffusion and drift of minority carriers in semiconductors for comparable capture and scattering mean free paths*," Phys. Rev., vol. 125, no. 5, pp. 1570–1576, Mar. 1, 1962.

114. W. Shockley and A. Goetzberger, "*The role of imperfections in semiconductor devices*," Proc. AIME, pp. 121–135, 1962.

115. K. Hubner and W. Shockley, "*Measurement of phonon scattering by a small angle grain boundary in silicon*," presented at the International Conference on Physics of Semiconductors, Institute of Physics and the Physical Society, Exeter, U.K., July 1962.

116. H. J. Queisser and W. Shockley, "*Diffusion-induced slip in silicon and the problem of dislocation distribution*," in Proc. 1st Berkeley Int. Materials Conf., Univ. California, Berkeley, 1963, pp. 781–789.

117. R. M. Scarlett, W. Shockley, and R. H. Haitz, "*Thermal instabilities and hot spots in junction transistors*," Physics of Failure in Electronics, M. F. Goldberg, Ed. Baltimore, MD: Cleaver-Hume, 1963, pp. 194–203.

118. R. M. Scarlett and W. Shockley, "*Secondary breakdown and hot spots in power transistors*," in IEEE Int. Convention Rec., 1963, pt. 3, pp. 3 13.

119. W. Shockley, "*Engineering challenges and human welfare*," Engineer's Week, 1963.

120. W. Shockley, D. R. Curran, and D. J. Koneval, "*Energy trapping and related studies of multiple electrode filter crystals*," in Proc. 17th Ann. Symp. Frequency Control, U.S. Army Electronics Research and Development Laboratory, Fort Monmouth, NJ, May 27–29, 1963, pp. 88–126.

121. R. H. Haitz, A. Goetzberger, R. M. Scarlett, and W. Shockley, "*Avalanche effects in silicon p-n junctions, I: Localized photomultiplication studies on microplasmas*," J. Appl. Phys., vol. 34, no. 6, pp. 1581–1590, June 1963.

122. W. Shockley, "*Scientific thinking and problems of growth*," Impact of Science: California and the Challenge of Growth, University of California. San Diego, CA: Univ. California Printing Department, 1963, pp. 90–103.

123. W. Shockley, *"Transistor history, applied research and science teaching,"* invited lecture, in Proc. 75th Anniversary Meeting Japanese Institution of Electrical Engineers, vol. 84–2, no. 905, 1963, pp. 147–158.

124. W. Shockley, H. J. Queisser, and W. W. Hooper, *"Charges on oxidized silicon surfaces,"* Phys. Rev. Lett, vol. 11, no. 11, pp. 489–490, Dec. 1, 1963.

125. W. Shockley and F. J. McDonald, *"Teaching scientific thinking at the high school level,"* Final Rep., Project S-090, U.S. Dept. Health, Education and Welfare, Office of Education Contract OE 4-10-216, School of Engineering and School of Education, Stanford Univ., Stanford, CA, 1964.

126. M. G. Buehler, W. Shockley, and G. L. Pearson, *"Hall measurements using Corbino-like current sources in thin circular disks,"* Appl. Phys. Lett, vol. 5, no. 11, pp. 228–229, Dec. 1, 1964.

127. W. Shockley and W. W. Hooper, *"The surface-controlled avalanche transistor,"* in Proc. Frontiers in Electronics Western Electronic Show and Convention, Los Angeles, CA, Aug. 25–28, 1964, pp. 1-3; see also Electronic Products, vol. 7, p. 68, 1964.

128. W. Shockley, W. W. Hooper, H. J. Queisser, and W. Schroen, *"Mobile electric charges on insulating oxides with application to oxide covered silicon p-n junctions,"* in Surface Science. Amsterdam, The Netherlands: North-Holland, 1964, vol. 2, pp. 277–287.

129. W. Shockley, lecture notes on *"Respect for the scientific nature of practical problems, and recognize the inadequacies in the 'law of excluded optimum,' as far as government agencies are concerned,"* in Proc. 3rd Navy Microelectronics Program Conf. Apr. 5, 1965. (Lecture notes taken and reconstructed by G. S. Szekely.)

130. R. Gereth and W. Shockley, *"Study of radiation damage by using field effect,"* Proc. IEEE, vol. 53, pp. 748–749, July 1965.

131. W. Shockley, J. A. Copeland, and R. P. James, *"The impedance field method of noise calculation in active semiconductor devices,"* in Quantum Theory of Atoms, Molecules, Solid State. New York: Academic, pp. 537–563.

132. W. Shockley, *"Articulated science teaching and balanced emphasis,"* IEEE Spectrum, vol. 3, no. 6, pp. 49–58, June 1966. Reprinted in Communication Concepts and Perspectives. Washington, DC: Spartan, 1967, pp. 153–179.

133. J. W. Allen, W. Shockley, and G. L. Pearson, *"Gunn domain dynamics,"* J. Appl. Phys., vol. 37, no. 8, pp. 3191–3195, July 1966.

134. Y. S. Chen, W. Shockley, and G. L. Pearson, *"Lattice vibration spectra of $GaAs_xP_{1-x}$ single crystals,"* Phys. Rev., vol. 151, no. 2, pp. 648–656, Nov. 11, 1966.

135. W. Shockley, *"Articulated science teaching and balanced emphasis,"* 37th Memorial Steinmetz Lecture, IEEE Spectrum, vol. 3, no. 6, pp. 49–58, 1966.

136. W. Shockley and R. P. James, *"A 'try simplest cases' development of forces on magnetic current,"* Bell Telephone Laboratories Technical Memo. with Abstract, Manuscript, and Diagrams, MM 67-25-1, Mar. 27, 1967.

137. W. Shockley, D. R. Curran, and D. J. Koneval, *"Trapped-energy modes in quartz filter crystals,"* J. Acoustic. Soc. Amer., vol. 41, no. 4 (pt. 2), pp. 981–993, Apr. 1967.

138. W. Shockley and R. P. James, *"'Try simplest cases' discovery of 'hidden momentum' forces on 'magnetic currents,'"* Phys. Rev. Lett, vol. 18, no. 20, CY501 (C), CY501 1-4, 2-4, 3-4, 4-4, pp. 876–879, May 15, 1965.

139. W. Shockley, responses to *"'Try simplest cases' discovery of 'hidden momentum' forces on 'magnetic currents,"* Ind. Res., p. 16, June 1967; *Bell Laboratories Rec,* p. 99, Mar. 1968; and *SCOPE—Stanford Electronics Laboratories,* vol. V, no. 2, Fall 1968.

140. S. M. Sze and W. Shockley, *"Unit-cube expression for space-charge resistance,"* Bell Syst. Tech. J., vol. XLVI, no. 5, pp. 837–842, May-June 1967.

141. W. Shockley, *"Hidden linear momentum" related to the E-term for a Dirac-electron wave packet in an electrical field,"* Phys. Rev. Lett, vol. 20, no. 7, p. 343 (C)-346, Feb. 12, 1968.

142. W. Shockley, P. D. Hurd, and F. J. McDonald, *"The con-servation of energy concept in ninth grade general science,"* U.S. Department of Health, Education and Welfare, Office of Education, Bureau of Research, Final Rep., Project no. OE 6-10-026, Feb. 1968, pp. i–ix, 1–84.

143. W. Shockley, *"A 'try simplest cases' resolution of the Abraham Minkowski controversy on electromagnetic momentum in matter,"* Proc. Nat. Acad. Sci., vol. 60, no. 3, pp. 807–813, July 1968.

144. W. Shockley and K. K. Thornber, *"The 'hidden momentum' equivalent to magnetic charges for a bound-state Dirac elec-tron,"* Phys. Lett, vol. 27A, no. 8, pp. 534–535, Sept. 9, 1968.

145. W. Shockley, *"Thinking about thinking improves thinking,"* IEEE Student J., pp. 11–16, Sept. 1968.

146. A. J. Kurtzig and W. Shockley, "A new direct measurement of the domain wall energy of the orthoferrites," *IEEE Trans. Magn.,* vol. MAG-4, pp. 426–430, Sept. 1968.

147. A. J. Kurtzig and W. Shockley, *"Measurement of the domain-wall energy of the orthoferrites,"* J. Appl. Phys., vol. 39, no. 12, pp. 5619–5630, Nov. 1968.

148. W. Shockley, *"S-ambiguity of Poynting's integral theorem eliminated by conceptual experiments with pulsed current dis-tributions,"* Phys. Lett, vol. 28A, no. 3, pp. 185–186, Nov. 18, 1968.

149. W. Shockley and K. K. Thornber, *"Hidden momentum for non-steady-state defined using a new mass-moment operator theorem for Dirac's equation,"* Phys. Lett, vol. 34A, no. 3, pp. 177–178, Feb. 22, 1971.

150. W. Shockley, *"Stark ladders for finite one-dimensional models of crystals,"* Phys. Rev. Lett, vol. 28, no. 6, pp. 349–352, Feb. 7, 1972.

151. W. Shockley, *"Three men who changed our world—25 years later,"* Bell Telephone Laboratories Rec, Dec. 1972.

152.   W. Shockley, *"The junction transistor,"* New Scientist, pp. 689–691, Dec. 21, 1972. Reprinted from *Bell Telephone Laboratories Rec,* pp. 379–381, Aug. 1951.

153.   W. Shockley, *"The invention of the transistor: An example of creative failure methodology,"* Solid State Devices, pp. 55–75, 1972.

154.   W. Shockley, *"The invention of the transistor—An example of creative failure methodology,"* National Bureau of Standards Special Publication 388, in *Proc. Conf. Public Need and the Role of the Inventor,* Monterey, CA, June 11–14, 1973, pp. 47–89.

155.   W. Shockley, *"The path to the conception of the junction transistor,"* IEEE Trans. Electron Devices, vol. ED-23, pp. 597–620, July 1976. Reprinted in *IEEE Trans. Electron Devices,* vol. ED-31, pp. 1523–1545, Nov. 1984.

156.   W. Shockley, *"Do dislocations hold technological promise?"* Solid State Technol, vol. 26, no. 1, pp. 75–78, Jan. 1983.

157.   W. Shockley, "Analysis suggesting major gene effects in the behavioral population genetics of Drosophila in phototaxis classification Mazes", Proc. NAS, Vol. 69 (1972), p. 1647A.

158.   W. Shockley, *"Mathematical models for assortative mating in American negro populations resulting in correlation between fractions of Caucasian ancestry,"* Proc. NAS, Vol. 70 (1973), p. 1619A.

159.   W. Shockley, *"Variance of Caucasian admixture in negro populations, pigmentation variability and IQ",* Proc. NAS, Vol. 70 (1973), p. 2180A.

160.   W. Shockley, *"American Lysenkoism in the National Academy of Sciences",* Congressional Record E7766, E7767, September 7, 1972.

161.   W. Shockley, *"Academy ruling unjust, resolution proposed again",* The Stanford Daily, pp. 2, October 17, 1972.

162.   W. Shockley, *"Transoprep acors, disrespect for truth in Academia",* The Stanford Daily, pp. 2, January 23, 1973.

163.   W. Shockley, *"Deviations from Hardy-Weinberg frequencies caused by assortative mating in hybrid populations",* Proc. NAS, vol. 70 (1973, pp. 732–736

164.   W. Shockley, *"Leeds affair is humanism gone Berserk",* The Times, pp. 1, March 16, 1973.

165.   W. Shockley, *"Leeds affair is humanism gone Berserk",* Manchester Union Leader, p. 43 and p. 52 March 28, 1973.

166.   W. Shockley, *"Confirmation of negro economic gain on whites and identification of dysgenic subpopulation by 'offset analysis'",* Proc. NAS, Vol. 70 (1973).

167.   W. Shockley, *"Sterilization – a thinking exercise",* The Stanford Daily, pp. 2, April 12, 1974.

168.   W. Shockley, *"Notes on the life and death of Tabby II",* Manchester Union Leader, p. 24 and p. 32, April 23, 1974.

169.   W. Shockley, *"The Relf Tragedy: facts and implication",* The Stanford Daily, pp. 2, October 24, 1973.

170. W. Shockley, *"The Relf Tragedy: Berserk Humanism or benevolent geno-cide"*, The Stanford Daily, pp. 2, October 23 1973.
171. W. Shockley, *"Is your genetic quality higher than an Amoeba's"*, Manchester Union Leader, p. 28 and p. 40, June 14, 1974.
172. W. Shockley, *"Crime and dysgenics,"* Skeptic, November-December 1974, pp. 50–51.
173. W. Shockley, *"The apple-of-Gods-eye obsession"*, The Humanist, Vol. 32, (1972), No. 1, pp. 16–17.

16.3   Lectures and Presentations

August 5, 1968	*"Transistor, Thinking about Scientific Problems, Thinking about Human Quality Problems"*
August 7–9, 1968	Three Lectures requested by Prof. D. Thomson for Latin American students, Stanford University
August 13–15, 1968	*"Hidden momentum forces on magnetic currents,"* TAC Meeting, Stanford University
August 28, 1968	*"A 'Try Simplest Cases' Resolution of the Abraham Minkowski controversy on electromagnetic momentum in matter,"* Theoretical Physics Seminar, Stanford University
October 14, 1968	*"Research on "Un-research" dogmatism of the environment-heredity uncertainty,"* SRI-RESA
October 28, 1968	*"Conceptual experiments, basic indeterminacy, and 'un-research' dogmatism,"* Autumn Meeting NAS, Caltech Pasadena, CA
November 1, 1968	*"Human quality problems and research taboos,"* Educational Record Bureau, New York, NY
November 14, 1968	*"Un-research thought blocking in electromagnetics and Eugenics,"* Fall Meeting of the Am. Soc. For Engineering education, Rickey's Hyatt House, School of Engineering
September 18, 1968	*"Research on "un-research dogmatism applied to the environment-heredity uncertainty,"* The Milwaukee Society, Milwaukee, Wisconsin
December 10, 1968	*"Human quality problems and research taboos,"* University of Bridgeport, Connecticut
January 20, 1969	*"Human quality problems and research taboos,"* Crothers Memorial, presentation for graduate engineering students
February 26, 1969	*"Human quality problems and research taboos,"* Actuarial Society, San Francisco
March 24, 1969	*"Thinking about thinking improves thinking,"* Bell System High School, Bell Telephone Laboratories, Murray Hill, NJ
March 31, 1969	Audio/Visual film at Hewlett-Packard Company on transistor for Palo Alto Diamond Anniversary

April 9 1969	*"Thinking about thinking improves thinking,"* FIA Association, Saratoga, CA
April 28, 1969	*"A polymolecular interpretation of growth rates of social problems,"* NAS Washington, D.C.
May 4, 1969	*"Thinking about thinking improves thinking,"* Escondido Group, Stanford University
May 9, 1969	*"Human quality problems and research taboos,"* IEEE Annual Meeting, Santa Barbara, CA
May 14, 1969	*"Human quality problems and research taboos,"* SEAS Symposium, New York, NY
June 27, 1969	*"Thinking about thinking improves thinking,"* MENSA, San Francisco
July 7, 1969	*"Human quality problems and research taboos,"* Dept. of Aerospace Engineering, Stanford University
July 29 1969	*"Human quality problems and research taboos,"* University of Colorado, Boulder, CO
August 17, 1969	Discussion of *"Human quality problems and research taboos,"* with Leonard L. Heston, Walter C. Alvarez, Robert Cancro and Robert E. Kuttner, Chicago, IL
August 21, 1969	*"Understanding and improving creativity,"* Am. Management Association, Chicago, IL
September 22, 1970	*"The environment-heredity uncertainty and research taboo,"* IEEE, Syracuse, New York
October 21, 1970	*"New methodology to reduce the environment-heredity uncertainty about dysgenics,"* Autumn Meeting NAS, Rice University, Houston, TX
November 4, 1970	*"Population pollution and the Speer syndrome,"* Dept. of *Electrical Engineering,* University of California, Santa Barbara, CA
November 17, 1970	*"The environment-heredity uncertainty and research taboo,"* Int. Symposium of Int. Society for Hybrid Microelectronics, Los Angeles, CA
November 30, 1970	*"The environment-heredity uncertainty and research taboo,"* Invitation from Engineering Students, Crothers Memorial, Stanford University
September 14, 1970	*"Invention of the transistor – an example of creative failure methodology,"* Am. Management Society, New York, New York
September 24, 1970	*"Thinking about thinking improves thinking"*, Am. Management Association, New York, New York
September 28, 1970	*"Transistor history,"* Prof. Padulo's class, Dept. of Electrical Engineering, Stanford University
April 7, 1971	*"Historical insights into research methodology,"* Dept. of Material Science, Stanford University (MS 239)

April 28, 1971	K.K. Thorbner, W. Shockley, *"Hidden momentum for non-steady state defined using a new mass-moment operator theorem for Dirac's equation"* Annual Spring Meeting, NAS, Washington, D.C.
January 14, 1971	*"Population pollution – the unmentionable threat to human resources,"* J. Smolensky's class, San Jose College and Isaac Newton Graham School, Mountain View, CA
April 28, 1971	*"Hardy-Weinberg law generalized to estimate hybrid variance for negro populations and reduce racial aspects of the environment-heredity uncertainty,"* Annual Meeting NAS, Washington, D.C
May 13, 1971	*"Dysgenics – the unthinkable, unmentionable threat,"* J. Smolensky's class, San Jose State College, L. C. Curtis School, Santa Clara, CA
May 26, 1971	*"Human quality problems and research tools,"* McClosky's graduate students, Dept. of Electrical Engineering, Stanford University
May 28–30, 1971	*"Human quality problems and research taboos,"* Lectures by request of students of Prof. Padulo's class, Stanford University
June 2, 1971	Debate, Prof. Watts and Prof. Shockley, Watt's Biology Class
June 24, 1971	*"Human quality problems and research taboos,"* Lectures requested by Dept. of Material
July 1, 1971	Science, Stanford University
September 7, 1971	*"Dysgenics – a social problem reality evaded by the illusion of infinite plasticity of human intelligence,"* in Symposium on Social Problems, Am. Psychological Association, Washington, D.C.
October 20, 1971	*"IQ-its significance and its geneticity evaluated by Las Vegas methods,"* Program in Human Biology, Stanford University
October 27, 1971	*"On the significance level for genetic dominance of IQ and on the 24-point difference between twins Gladys and Helen,"* NAS, Washington, D.C.
November 6, 1971	*"Human quality problems and research taboos,"* Bechtel International Center, Stanford University
November 9. 1971	*"Population pollution – the unthinkable dysgenic threat,"* Johnson Club, Los Angeles, CA
November 19, 1971	*"Research Taboos on geneticity of IQ,"* Ingenieurs Civils de France, West Coast Section, San Francisco, CA
November 22, 1971	*"IQ-its significance and its geneticity evaluated by Las Vegas method,"* Sacramento State College, Sacramento, CA
November 29, 1971	*"IQ-its significance and its geneticity,"* Hammarskjold House, University of California at Davis, Davis, CA
February 25, 1972	*"Research taboos on geneticity of IQ,"* Santa Clara Valley Engineers Council, San Jose Hyatt House, San Jose, CA

March 1, 1972	*"Research taboos on geneticity of IQ,"* Professional Journalism Fellows Seminar, Dept. of Communication, Stanford University
March 6, 1972	*"Research taboos on geneticity of IQ,"* Grove House, Stanford University
March 22, 1972	*"Environment-heredity uncertainty and research taboos,"* Wilbur Wright College, Chicago, IL
March 23, 1972	*"Research taboos and geneticity of IQ,"* MIT Alumni Pennsylvania, Bethlehem, Pennsylvania
March 24, 1972	*"Environment-heredity uncertainty and research taboos,"* The Stoics Summit, New Jersey, NJ
April 21, 1972	*"IQ and Genetics,"* Dr. Charlton Social Psychiatry Seminar
April 26, 1972	*"Analysis suggesting major gene defects in the behavioral population genetics of drosophila in phototoxic classification mazes,"* Contributing paper NAS, Washington, D.C.
May 5, 1972	*"IQ and genetics,"* Dr. Charlton Social Psychiatry Seminar, Stanford University
May 9, 1972	*"Research taboos on geneticity of IQ,"* Graduate School of Business, Stanford University
May 16, 1972	*"Research taboos on geneticity of IQ,"* Canada College, Redwood City, CA
May 17, 1972	*"Research taboos on geneticity of IQ,"* Grove House, Stanford University
May 21, 1972	*"Research taboos on geneticity of IQ,"* Episcopal Trinity Parish, Menlo Park, CA
June2, 1972	*"Research taboos on geneticity of IQ,"* UCLA Los Angeles, CA
July 5, 1972	*"Estimates of level of significance for genetic dominance of IQ for Caucasian,"* Statistics Dep. of Stanford University
August 3, 1972	*"80% Geneticity of IQ,"* Stanford Hospital House Staff Wives' Club, Stanford, CA
October 9, 1972	*"Estimates of level of significance for genetic dominance of IQ for Caucasians,"* Invited lecture for the Air Force Squadron Reserve, Lockheed Corp., Palo Alto
October 16, 1972	*"Mathematical models for Assortative mating in American Negro populations resulting in correlation between fraction of Caucasian ancestry,"* NAS, Fall Meeting, Washington, D.C.
October 29, 1972	*"Human quality problems,"* Beth Am. Jewish Synagogue, Los Altos, CA
December 7, 1972	*"80% geneticity of IQ,"* Monterey Peninsula College, Monterey, CA
January 7, 1973	*"Human quality problems and research taboos,"* Palo Alto Unitarian Church, Palo Alto, CA

January 11, 1973	*"Human quality problems and research taboos,"* Annual meeting of the Unitarian Universalist Ministers, West Coast Association, Los Angeles, CA
January 14, 1973	*"Human quality problems and research taboos,"* Emerson Unitarian Church, Canoga Park, CA
January 21, 1973	*"Human quality problems and research taboos,"* Palo Alto Unitarian Church, Palo Alto, CA
January 22, 1973	*"80% at 2,000 IQ geneticity,"* Skilling Auditorium, Stanford University
January 23, 1973	*"The nature of heritability of IQ and its social, economic and political implications,"* Psychology Department Colloquia, Memorial Auditorium, Stanford University
February, 12, 1973	*"Estimation of proportions in racial admixture,"* Invited lecture, Galton Laboratory, University College London, England
February 16, 1973	*"The 80% at 2,000 IQ geneticity Assertion,"* British Science Writer Association, Hotel Russell, London, England
April 24, 1973	*"Variance of Caucasian admixture in Negro population, pigmentation, variability and IQ,"* NAS, Spring meeting, Washington, D.C.
June 14, 1973	*"The voluntary sterilization bonus plan – an answer to population variability and IQ,"* Lake Merritt Breakfast Club, Garden Centre Oakland, CA
August 24, 1973	*"Eugenics, Dysgenics, and freedom of the press – the lesson of Nazi history anticipated in our First Amendment,"* Invited Lecture, Presbyterian Church, Birmingham, Alabama
September 17, 1973	*"Berserk humanism or benevolent genocide?",* KIWANIS club of the Peninsula
September 19, 1973	*"Interpretation of β degradation by analysis of channel characteristics of Field Effect Transistor,"* Solid State Industrial Affiliate Review (with Albert Wang)
October 1, 1973	*"Human quality problems and research taboos,"* Congregational Presbyterian and Unitarian Churches in Palo Alto, CA
October 9, 1973	*"The invention of the Transistor – an example of creative failure methodology,"* Power Engineering Society, Engineering Club San Francisco, CA
October 10, 1973	*"80% (at 2,000) IQ geneticity assertion",* Freshman Seminar (with Prof. H.O. Fuchs), Stanford University
October 11, 1973	*"IQ and its geneticity",* Napa Valley College, Napa, CA
October 17, 1973	*"IQ and its geneticity",* debate with Cavalli Sforza, Stanford University
November 8, 1973	*"80% (at 2,000) IQ geneticity assertion",* University of Washington, Seattle, WA

November 20, 1973	*"80% (at 2,000) IQ geneticity assertion"*, Staten Island Community College, New York, with Frances Welsing
November 23, 1973	*"The invention of the Transistor – an example of creative failure methodology "*, presentation for Motorola Science Advisory Board, Phoenix, AZ
December 4, 1973	"IQ differences, heredity and dysgenics", Princeton University, Princeton, N.J.
January 19, 1974	*"80% (at 2,000) IQ geneticity assertion"*, Cold Spring Harbor High School
March 5 1974	*"80% (at 2,000) IQ geneticity assertion"*, University of Georgia, Athens, GA
March 13, 1974	*"80% (at 2,000) IQ geneticity assertion"*, Awalt High School, Mountain View, CA
April 9, 1974	*"The moral obligation to diagnose the American negro tragedy of statistical IQ deficit",* The New York University, New York, NY
April 16, 1974	*"Shockley Reminiscence,"* Dept. of Material Science, Stanford University
April 15, 1974	*"Resolved: Society should diagnose and treat tragic racial IQ inferiorities",* Yale University, New Haven, CT, debate with W. Rusher
April 30, 1974	*"Invention of the transistor,"* Dept. of Material Science, Stanford University
May 6, 1974	*"The invention of the Transistor – an example of creative failure methodology,"* Virginia Polytechnical University, Blacksburg, Virginia
May 14, 1974	*"Grain boundaries and dislocations,"* Dept. of Material Science, Stanford University
May 16, 1974	*"Human quality problems and research taboos,"* Lecture for The Barristers of the Los Angeles, Los Angeles, CA
May 17, 1974	*"The American tragedy of negro statistical and mental inferiority",* The Kenna Club, University of Santa Clara, CA
May 23, 1974	*"80% (at 2,000) IQ geneticity assertion"*, University of Maryland, College Park, MD
May 28, 1974	*"Magnetic bubbles,"* Dept. of Material Science, Stanford University
June 4, 1974	*"Dislocation pipe diffusion in semiconductors,"* Dept. of Material Science, Stanford University
September 15, 1974	*"The moral obligation to diagnose the American negro tragedy of statistical IQ deficit",* Case Western University, Cleveland, OH, debate with Roy Innis
October 14, 1974	*"Delinquency and crime,"* Seminar at Metropolitan Club, Washington, D.C.
November 21, 1974	*"Science and pseudoscience,"* Freshman seminar, Stanford University

December 9, 1974	*"Human quality problems and research taboos,"* Union Club New York City, New York, NY
December 17, 1974	*"Humanitarianism gone Berserk,"* Albany Rotary Club, Albany, NY
February 5, 1975	*"The moral obligation to diagnose the American negro tragedy of statistical IQ deficit,"* University of Virginia, Charlottesville, VA
February 6, 1975	*"Resolved: Society should diagnose and treat tragic racial IQ inferiorities,"* University of Illinois, Chicago, IL
February 14, 1975	*"The moral obligation to diagnose the American negro tragedy of statistical IQ deficit,"* University of San Fernando, Law School, (now UWLA) Inglewood, CA
February 22, 1975	*"Human quality problems and research taboos,"* presentation for United Republican of California, Sacramento, CA
March 3, 1975	*"The moral obligation to diagnose the American negro tragedy of statistical IQ deficit,"* University of Nebraska, Omaha, NE, debate with Roy Innis

16.4 Radio and Television Appearances

September 11, 1971	WCAU (CBS) Radio Philadelphia, Moderator: Dom Quinn
September 12, 1971	WMCA Radio New York, Moderator: Jeffory St. John
September 13, 1971	CBS Radio, Moderator: William Ronald
September 19, 1971	KNXT-TV Los Angeles, Moderator: Jacques Truman
September 20, 1971	CBC-TV Toronto, Canada, *"Front Page Challenge"*
October 9, 1971	KGO-TV San Francisco, *"On the spot"*
November 23, 1971	KOVR-TV Sacramento, Moderator: Chet Trouten
December 5, 1971	KCBS Radio San Francisco, "In depth," Moderator: Don Mozley
December 5, 1971	KZAP Radio, Sacramento, Moderator: "Ace"
December 13, 1971	KFBK Radio Sacramento, Moderator: Al Hooker
February 2, 1972	KZSU Radio, Stanford University
February 13, 1972	KRON-TV San Francisco, *"Speak out,"* Moderator: Dave Valentine
May 15, 1972	ABC-TV Hollywood, *"A.M. Show,"* Moderator: Ralph Story
June 24, 1972	NBC TV Los Angeles *"IQ and Race"*
October 19, 1972	Thames TV, *"Something to say,"* debate with Ashely Montagu.
December 6, 1972	KBKT-TV, KBSC-TV, KBC-TV, WKBF-TV, WKBS-TV *"The Lou Gordon show"*
December 8, 1972	KABC-Radio, "Interview with Paul Mayeda, John Babcock, Bob Arthur and Michael Jackson"

December 12, 1972	WISN-TV *"Interview with Bob Viverito"*
February 18, 1973	WKBD-TV *"The Lou Gordon show,"* with Dr. Scarr-Salapatek
February 25, 1973	BBC-TV *"David Frost show,"* London, England
March 15, 1973	KSTP-TV *"Charlie McCarthy show"*
July 6, 1973	Stanford Educational TV
August 13, 1973	KQED-TV *"Say it loud with Donald Warden"*
August 31, 1973	KRON-TV *"Interview with Dave Valentine"*
September 7, 1973	KRON-TV *"Interview with Dave Valentine"*
September 25, 1973	KHJ-TV *"Philbin and Co."*
October 29, 1973	WNEW-TV *"David Susskind show"*
November 26, 1973	WCBS-TV *"Today show"* with Roy Innis
November 27, 1973	NBC-TV *"Take it from here,"* with Dr. Frances Welsing
December 6, 1973	WMCA Radio *"Bob Grant show"*
December 10, 1973	CBS-TV *"Tomorrow show,"* with Roy Innis
December 14, 1973	KT-TV *"Let's Rap,"* with Roy Innis
January 3, 1974	WAAB Worcester, Massachusetts
January 10, 1974	KPIX *"Community News Conference"*
January 16, 1974	WWRL New York City, interview with Alvin Poussant
January 22, 1974	WMCA Radio
February 5, 1974	WNET-TV *"Black Journal,"* debate with Dr. Frances Welsing
April 7, 1974	CBS *"60 min,"* with Mike Wallace
April 13, 1974	"KLAT's Conference Line," Salt Lake City
May 10, 1974	WLAC-TV *"Stanley Siegel show"*
May 9, 1974	WCBS-TV, interview with Sol Panitz
May 10, 1974	WRN6 Radio Atlanta
May 24, 1974	WCBS-TV *"Pat Collins show,"* with Roy Innis
June 11, 1974	NBC-TV "Carol Simpson 10"
June 30, 1974	RELEVANCE, Radio Talk show
July 21, 1974	KQED-TV *"Firing Line,"* with W.F. Buckley
August 30, 1974	KGBS Los Angeles *"Mickey and Teddi Radio show"*
August 30, 1974	KTTV Metro News Nevada, interview with Larry Attebery
September 16, 1974	WERE Radio, *"Disruption at Case Western Reserve University"*
September 16, 1974	WKYC-TV Cleveland *"Sunday Magazine"*
October 17, 1974	WRR Radio, interview with Bill Jameson and Guy Gibson
October 26, 1974	KKTV-TV Los Angeles, *"Meet David Sachs M.D."*
March 17, 1975	KGOP-TV Los Angeles, *"Both sides now"*
April 25, 1975	WJW-TV Cleveland *"Discussion with Sidney Andorn on academic freedom"*
April 25, 1975	WEWS-TV Cleveland, interview with Dorothy Fuldheim

Name Index

Subject Index